Biomedical Technology Assessment: The 3Q Method

Biomedical Technology Assessment: The 3Q Method
Phillip Weinfurt

ISBN: 978-3-031-00513-8   paperback
ISBN: 978-3-031-01641-7   ebook

DOI 10.1007/978-3-031-01641-7

A Publication in the Springer series
SYNTHESIS LECTURES ON BIOMEDICAL ENGINEERING

Lecture #33
Series Editor: John D. Enderle, *University of Connecticut*
Series ISSN
Synthesis Lectures on Biomedical Engineering
Print 1947-945X   Electronic 1947-9468

Synthesis Lectures on Biomedical Engineering

Editor

John D. Enderle, *University of Connecticut*

Lectures in Biomedical Engineering will be comprised of 75- to 150-page publications on advanced and state-of-the-art topics that spans the field of biomedical engineering, from the atom and molecule to large diagnostic equipment. Each lecture covers, for that topic, the fundamental principles in a unified manner, develops underlying concepts needed for sequential material, and progresses to more advanced topics. Computer software and multimedia, when appropriate and available, is included for simulation, computation, visualization and design. The authors selected to write the lectures are leading experts on the subject who have extensive background in theory, application and design.

The series is designed to meet the demands of the 21st century technology and the rapid advancements in the all-encompassing field of biomedical engineering that includes biochemical, biomaterials, biomechanics, bioinstrumentation, physiological modeling, biosignal processing, bioinformatics, biocomplexity, medical and molecular imaging, rehabilitation engineering, biomimetic nano-electrokinetics, biosensors, biotechnology, clinical engineering, biomedical devices, drug discovery and delivery systems, tissue engineering, proteomics, functional genomics, molecular and cellular engineering.

Biomedical Technology Assessment: The 3Q Method
Phillip Weinfurt
2010

Strategic Health Technology Incorporation
Binseng Wang
2009

Phonocardiography Signal Processing
Abbas K. Abbas, Rasha Bassam
2009

Introduction to Biomedical Engineering: Biomechanics and Bioelectricity - Part II
Douglas A. Christensen
2009

iv

Introduction to Biomedical Engineering: Biomechanics and Bioelectricity - Part I
Douglas A. Christensen
2009

Landmarking and Segmentation of 3D CT Images
Shantanu Banik, Rangaraj M. Rangayyan, Graham S. Boag
2009

Basic Feedback Controls in Biomedicine
Charles S. Lessard
2009

Understanding Atrial Fibrillation: The Signal Processing Contribution, Part I
Luca Mainardi, Leif Sörnmo, Sergio Cerutti
2008

Understanding Atrial Fibrillation: The Signal Processing Contribution, Part II
Luca Mainardi, Leif Sörnmo, Sergio Cerutti
2008

Introductory Medical Imaging
A. A. Bharath
2008

Lung Sounds: An Advanced Signal Processing Perspective
Leontios J. Hadjileontiadis
2008

An Outline of Informational Genetics
Gérard Battail
2008

Neural Interfacing: Forging the Human-Machine Connection
Thomas D. Coates, Jr.
2008

Quantitative Neurophysiology
Joseph V. Tranquillo
2008

Tremor: From Pathogenesis to Treatment
Giuliana Grimaldi, Mario Manto
2008

Introduction to Continuum Biomechanics
Kyriacos A. Athanasiou, Roman M. Natoli
2008

The Effects of Hypergravity and Microgravity on Biomedical Experiments
Thais Russomano, Gustavo Dalmarco, Felipe Prehn Falcão
2008

A Biosystems Approach to Industrial Patient Monitoring and Diagnostic Devices
Gail Baura
2008

Multimodal Imaging in Neurology: Special Focus on MRI Applications and MEG
Hans-Peter Müller, Jan Kassubek
2007

Estimation of Cortical Connectivity in Humans: Advanced Signal Processing Techniques
Laura Astolfi, Fabio Babiloni
2007

Brain–Machine Interface Engineering
Justin C. Sanchez, José C. Principe
2007

Introduction to Statistics for Biomedical Engineers
Kristina M. Ropella
2007

Capstone Design Courses: Producing Industry-Ready Biomedical Engineers
Jay R. Goldberg
2007

BioNanotechnology
Elisabeth S. Papazoglou, Aravind Parthasarathy
2007

Bioinstrumentation
John D. Enderle
2006

Fundamentals of Respiratory Sounds and Analysis
Zahra Moussavi
2006

Advanced Probability Theory for Biomedical Engineers
John D. Enderle, David C. Farden, Daniel J. Krause
2006

Intermediate Probability Theory for Biomedical Engineers
John D. Enderle, David C. Farden, Daniel J. Krause
2006

Basic Probability Theory for Biomedical Engineers
John D. Enderle, David C. Farden, Daniel J. Krause
2006

Sensory Organ Replacement and Repair
Gerald E. Miller
2006

Signal Processing of Random Physiological Signals
Charles S. Lessard
2006

Image and Signal Processing for Networked E-Health Applications
Ilias G. Maglogiannis, Kostas Karpouzis, Manolis Wallace
2006

Biomedical Technology Assessment: The 3Q Method

Phillip Weinfurt
Marquette University

SYNTHESIS LECTURES ON BIOMEDICAL ENGINEERING #33

ABSTRACT

Evaluating biomedical technology poses a significant challenge in light of the complexity and rate of introduction in today's healthcare delivery system. Successful evaluation requires an integration of clinical medicine, science, finance, and market analysis. Little guidance, however, exists for those who must conduct comprehensive technology evaluations. The 3Q Method meets these present day needs. The 3Q Method is organized around 3 key questions dealing with 1) clinical and scientific basis, 2) financial fit and 3) strategic and expertise fit. Both healthcare providers (e.g., hospitals) and medical industry providers can use the Method to evaluate medical devices, information systems and work processes from their own perspectives. The book describes the 3Q Method in detail and provides additional suggestions for optimal presentation and report preparation.

KEYWORDS

3Q Method, biomedical technology, biomedical technology assessment, healthcare technology assessment, hospital technology evaluation, medical technology evaluation methodology, clinical technology assessment, healthcare technology business assessment

Contents

Acknowledgments xi

Preface xiii

1 Introduction 1

    1.1 Scenario #1 1

    1.2 Scenario #2 1

    1.3 Scenario #3 1

    1.4 Context for Using the Method 2

    1.5 Types of Technology Suitable for Evaluation using the 3Q Method 3

2 Question #1: Is It Real? 5

    2.1 Does it Solve a Real Clinical Problem? 5

        2.1.1 What is the Clinical Problem?   6

        2.1.2 What is the Projected Clinical Impact?   6

        2.1.3 Does it Create New Problems?   7

    2.2 Is it Clinically Tested? 7

        2.2.1 Categories of Clinical Studies   8

        2.2.2 Clinical Limits Tested?   25

        2.2.3 Clinical Usability   26

        2.2.4 Training Complexity?   27

    2.3 Is it Based Upon Real Scientific Principles? 27

3 Question #2: Can We Win? 29

    3.1 Can We Win (Company)? 30

        3.1.1 Is it A Strategic Fit for Us?   30

        3.1.2 Is it Within our Expertise?   31

3.1.3 Can we be Competitive?   32

3.1.4 Do we Have Worldwide Distribution Channels?   32

3.1.5 Is the Customer Adoption Curve Acceptable?   33

3.1.6 What are the Regulatory Requirements?   34

3.2   Can we Win (Hospital)? 34

3.2.1 Is it a Strategic Fit for Us?   35

3.2.2 What is the Vendor Capability?   36

3.2.3 Can we be Competitive?   36

4   **Question #3: Is It Worth It?** 39

4.1   Company Setting 40

4.2   Hospital Setting 41

4.3   Conclusion 46

5   **3Q Case Study Example – Pershing Medical Company** 49

5.1   Pershing Medical Company (PMC) Background 49

5.2   Current Situation 50

5.3   Is It Real? 50

5.4   Can We Win? 56

5.5   Is It Worth It? 60

A   **Health Care Technology Assessment Sample Class Syllabus** 65

B   **How Do Hospitals and Clinicians Get Paid?** 67

C   **Technology Assessment PowerPoint Report Guidelines** 73

D   **Class Report Scenario Example** 75

E   **Four-Blocker Slide Templates for 3Q Reports** 77

Author's Biography 85

Acknowledgments

This book is the result of five years of creating and modifying my class notes for the course I have been teaching. It also is the result of a number of important people in my life who have supported me and made many contributions to this book. I would first like to thank my wife, Mary, who has persevered through this long writing process.

I am especially thankful for the many contributions from my son, Kevin, who is allied with me in the healthcare field in his research work at Duke University. His knowledge of healthcare statistical methods and applications, as well as his well-developed sense of clear writing has helped me immeasurably. Finally, his encouragement during this process was instrumental in my completing this work.

I thank those that reviewed this work during the draft stages. Thank you Colleen Jensen, DVM, my daughter, also allied with me in the healthcare field and Steve Emery, MS, FCCM, my good friend and healthcare colleague.

Getting me started in this endeavor was my colleague and the Director of our Healthcare Technologies Management Program at Marquette University, Jay Goldberg, Ph.D., PE.

I am indebted to my long-time colleague and friend, Patricia Brigman, who contributed significantly to the course content and format, as well as acting as consultant to my students over the past four years. Her expertise as an executive coach has added a unique dimension to my course for which I am extremely grateful.

I would also like to thank my good friend and long-time colleague, Gary Earl, Ph.D., who has provided a wide industrial perspective to this book and to my class. Additionally, Gary, along with other colleagues at GE Healthcare, have acted as industry experts for my students over the past five years. I thank you for your generous time and continued support.

I would like to thank all the students over the last five years who contributed in one way or another to the contents of this book. Their questions led me to add, embellish, discard and clarify this material.

Special thanks and appreciation goes to my colleague and draft editor, Julie Hoesly. Her professional writing talents and patience were invaluable to the completion of this project.

Greatly assisting me in this work at Morgan & Claypool Publishers was Joel Claypool. My thanks to him and all his staff for their patience and support.

Phillip Weinfurt, Ph.D.
February 2010

Preface

The 3 Question (3Q) medical technology assessment method is something I created over the last five years while teaching a medical technology assessment graduate course at Marquette University. Most of my students have an undergraduate degree in Biomedical Engineering. However, the course is intended for any student desiring a background in healthcare technology assessment, in either a scientific, clinical or business setting.

In preparing for the first semester of teaching, I found little cohesive material on the span of subjects I wanted to cover. Much of the existing material on medical technology assessment was created years ago, and was limited in scope and relevancy to today's standards. Consequently, I began to draw upon my experience and develop a methodology, which I named the 3 Question (3Q) medical technology assessment method. Many years as an industry healthcare professional with two device companies had generated a proven process that I wanted to document and share.

The course contained in this book is a semester-based course, with 16 three-hour classes held once a week. An optimal class size is six to eight students. The class format is typically lecture followed by discussion. I present new material in much the same order as the 3Q diagram, with the use of several cases to illustrate the main points. The last portion of the class is open to general discussion, but frequently I allocate the last half hour for the report teams to work on their final 3Q reports. This allows for individual questions about any process, presentation or content.

Several times during the semester I take the class on a field trip to places such as an ICU, a medical school simulation laboratory and a medical device company. These different settings provide additional input to the course content. For added perspective, I invite a venture capitalist to guest lecture on the 3Q Method as well. Throughout this course, it's my goal to transform these students into future employees who can make a positive impact with their knowledge and analysis.

To that end, students are given the opportunity for practical application of the 3Q Method through the use of case studies. Case studies usually rotate between company and hospital scenarios to expose students to the difference in the 3Q process based on its setting. Company case studies involve a scenario in which the students are company employees who are asked to investigate a new technology and recommend to management whether the company should pursue it further. Hospital case studies involve a scenario in which the students are employees of a hospital working in the Technology Assessment Office. The students' task is to assess a new technology for possible purchase and present their recommendation to a hospital review committee made up of administrators and clinicians.

Obtaining case studies can be problematic since they require: a) company or hospital market information and, b) company and hospital financial information that may be confidential. However,

I encourage resourcefulness in my students to use the Internet or personal contact to generate case study material. I have resisted canned case studies for this reason.

Over the course of the semester, four to five major case study analysis reports are assigned to teams of two to three students. The students have at least two weeks to complete their reports, which are presented by the teams via PowerPoint. These reports are graded on content, presentation and slide design. Several healthcare professionals attend the presentation sessions to offer their feedback. This practice has been very effective in enhancing the students' experience with real-life context and contact. This is an important objective of the course. A sample course syllabus is provided in the Appendix.

By the end of the course, students should have gained a clear understanding of how to apply the 3Q Method in a scientific, clinical or business setting. In doing so, they will also acquire the following:

- Critical assessment of medical journal articles

- Application of Bland Altman analysis

- Understanding of two-arm clinical studies

- Facility with incremental five-year NPV financial analysis

- Facility with market analysis, including market share and market growth

- Familiarity with clinical measurement technologies

- Familiarity with clinical therapeutic technologies

- Familiarity with clinical informatics technologies

- Knowledge of the healthcare reimbursement system

- Understanding of professional presentation skills

- Understanding of professional PowerPoint design

- Communication skills with healthcare clinicians and administrators

- Communication skills with industry management

- Knowledge and comfort in the business environment

Critical thinking skills must be used in conjunction with the 3Q Method. Throughout this course, I stress the following:

1. Ask questions; be willing to wonder

2. Define your problem correctly

3. Examine the evidence

4. Analyze assumptions and biases

5. Avoid emotional reasoning

6. Don't oversimplify

7. Consider other interpretations

The structure of this book follows the 3Q Process Map on page x. Other reference material and examples are found in the Appendix. It's my hope that these contents will serve as a valuable reference tool for all individuals faced with the daunting task of making decisions on new healthcare technology.

Phillip Weinfurt, Ph.D.
February 2010

CHAPTER 1

Introduction

1.1 SCENARIO #1

Two weeks ago you received a telephone call from an inventor (Ivan) who would like to see if your enterprise would be interested in his new medical technology. He tells you that his device will measure blood pressure without the arm cuff that is commonly used today. It is small, unobtrusive, inexpensive and very accurate. The device has been tested on a number of individuals against similar instruments, with impressive results. It gives a continuous reading of heart rate, systolic, diastolic and mean blood pressure, and displays a beat-by -beat blood pressure waveform – all features that present cuff technology does not provide. Ivan asks you to review some product data he is going to send you and respond back in two weeks with your interest level. Your manager asks you to assess this new technology and lets you know that she is especially skeptical of the scientific principles behind this device and the validity of the clinical tests.

1.2 SCENARIO #2

A week ago, you received an e-mail from a physician (Phil) describing a product that he has been working on for the last two years. Phil would like to know if your company would be interested in it. It's a software program to aid clinicians in making the best choice of antibiotics for intensive care patients. Phil tells you that this decision is important because choosing an ineffective antibiotic will give the bacteria additional time to multiply. Phil believes that this software will be ideally suited for your company's patient monitoring product line, and it could provide your company with a competitive edge. He asks you to look over his software specifications and present them to the appropriate principals at your company. He will call you in a week to see if your company is interested in further follow-up. You inform your manager about this product idea, and she asks you to assess if this, indeed, would be synergistic with the company's current and future patient monitoring product strategy.

1.3 SCENARIO #3

You are working for a metropolitan hospital. Today, the Chief Technology Officer (CTO) of your hospital comes to you and tells you that a company that designs and manufactures pulse oximeters has brought in a new, and greatly improved, pulse oximeter. The company said that its 510(k) submission has FDA approval and that the pulse oximeter is presently being tested in five major hospitals with great success. Its major patient care benefit is that it is extremely resistant to patient movement artifact while providing accurate heart rate and arterial oxygen saturation during patient movement. You know that this new product could potentially benefit patients in the emergency department, intensive care units, surgery and the medical/surgical floors. Your CTO likes the technology of artifact resistance, but is concerned about the cost of replacing all of the present hospital pulse oximeters. He would like you to analyze this new pulse oximeter from a clinical and financial standpoint.

Note that in each scenario, your "boss" is concerned about a different aspect of the technology. Do you know how to address *all* of these aspects in a methodical, objective fashion? After reading this book, you will be able to complete an analysis of new medical technology, create a high quality, professional report and make a compelling presentation to your manager. This is the purpose of the 3Q Method.

The 3Q Method provides an analysis framework, which contains the most important assessment elements for arriving at a concise and high quality evaluation of a particular medical technology. The Method applies various assessment tools to a range of medical technologies to determine the clinical utility and the business feasibility of these technologies.

The 3Q Method is comprised of asking three major questions:

1. **Is it real?**

2. **Can we win?**

3. **Is it worth it?**

Contained in these top-level questions are sub-questions, which lead the reader through a detailed analysis of the subject technology. In this fashion, the 3Q Method becomes a sequential checklist of questions. The "*Is it real?*" question covers the scientific and clinical viability of the technology. The "*Is it worth it?*" and "*Can we win?*" questions cover the business aspects of the technology. Hence, the user arrives at a comprehensive scientific, clinical and business analysis of the technology. The topics covered in the 3Q Method will, in the end, produce a suitable report format for presentation to management.

1.4   CONTEXT FOR USING THE METHOD

The 3Q Method is designed to address two institutions – companies and hospitals. The methodology is applicable to both. By "company," I mean any size enterprise that designs, manufactures and/or markets medical devices or systems for the worldwide market. Start-up companies are included, but they are a special case, i.e., they may choose to market their products through other companies, or they may choose to sell their technology to another company. From the medical company prospective, the 3Q Method is used to evaluate a particular technology for the purpose of integrating this technology into the company's product line(s). It is not intended as a full due diligence analysis[1] for the purpose

---

[1] A method used to analyze a potential investment; it is an in-depth study of markets, products, company strengths and weaknesses, etc.

of buying a company, although the 3Q Method can be used as a starting point for a complete due diligence analysis.

The term "hospitals" used throughout this book refers to any care delivery institution or system, regardless of size, location or specialty. From a hospital's prospective, the 3Q Method would be used to arrive at an objective comparison of vendor products prior to the acquisition of a particular medical device or system.

1.5   TYPES OF TECHNOLOGY SUITABLE FOR EVALUATION USING THE 3Q METHOD

The 3Q Method is applicable to various types of medical technologies:

Measurement devices — these are typically used in acute care hospitals and clinics to measure physiological variables in the body for the purposes of determining or monitoring the effectiveness of medical therapies. Examples are the following: patient monitoring systems, body chemistry lab systems, screening systems and imaging systems.

Therapeutic devices — these are devices that assist and maintain the function of the body. Examples are the following: IV pumps, ventricular assist devices, dialysis systems and oxygen delivery devices.

Clinical information systems — these are a wide range of systems that process and store patient information. Examples are the following: acute care information systems, clinic information systems, electronic medical records and decision support systems.

Not all aspects of the 3Q Method will be necessary to complete for every technology evaluation. For example, an existing device being evaluated for a new customer market probably will not need the "Is it real?" analysis. However, since this is difficult to determine a priori, the entire 3Q Method should be considered for any technology assessment.

The Method is not intended for the analysis of pharmaceuticals. Pharmaceutical technology development, science, clinical testing and marketing are very different from devices and information systems and, as such, demand a unique analysis method. Many publications on the pharmaceutical industry are referenced throughout the book.

C H A P T E R   2

Question #1: Is It Real?

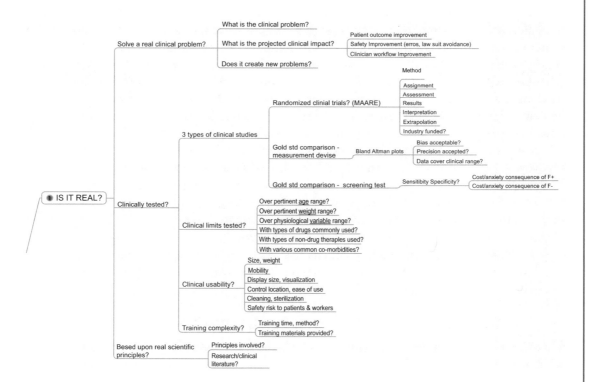

Figure 2.1: 3Q Method Map – Question #1: Is It Real?

This section examines both the scientific basis and the clinical assessment of the technology. To begin this analysis, these topics will be dissected into more specific questions, as illustrated in the diagram on Figure 2.1.

2.1   DOES IT SOLVE A REAL CLINICAL PROBLEM?

This subtopic breaks down into three questions:

1. What is the clinical problem?

2. What is the projected clinical impact?

3. Does it create new problems?

Each of these questions will be covered.

2.1.1   WHAT IS THE CLINICAL PROBLEM?

It is vitally important to clearly identify the clinical problem that the proposed technology will solve. The clinical problem should be defined and refined through documented discussions with the clinical users of the proposed technology (e.g., doctors, nurses, technicians, etc.). This problem definition phase should not include any possible solutions or introduction of the new technology. Clearly, defining the clinical problem will make it easier to determine if the features of the new technology will indeed solve the problem. Furthermore, careful definition of the problem should lead naturally to the selection of measurable variables that should improve as a result of introducing the new technology. For that reason, it is beneficial to define the problem in quantitative terms if possible, using clinical outcome variables such as mortality or hospital length of stay (LOS). This allows the clinical effectiveness to be more easily quantified when one gets to that point in the analysis. Objectively, defining the clinical problem will lead to an accurate assessment process. For example, if the clinical problem is defined in terms of reducing LOS, mortality, medication cost, infections, number of lab tests, etc., it will be in the "language" of the hospital clinicians and administrators and be immediately understood. Using more qualitative outcomes for analysis such as "greater accuracy," "better care," and "improved workflow" are too subjective and general to be of use to the stakeholders. The direct, quantitative effect on patients, care providers, administrators, workflow, clinical protocols, etc., must be sought out and clearly presented in the 3Q report. This is your important first challenge at the start of your analysis.

2.1.2   WHAT IS THE PROJECTED CLINICAL IMPACT?

While the technology under investigation may be based upon scientific principles and may solve a real clinical problem, it may have questionable impact on your specific healthcare delivery system. This needs to be researched and detailed in the report to ensure a comprehensive approach to the decision-making process. Four questions are helpful in determining clinical impact:

1. Is there a quantitative improvement in patient outcome (e.g., LOS reduction, reduction of hospital acquired infections, etc.)?

2. Is there an impact on patient safety (e.g., lawsuit avoidance, reduction in errors)?

3. Is there an improvement in clinical workflow, (e.g., faster turnaround for lab blood results)?

4. How many patients will this affect per year?

Answers to these questions should be expressed using statistical data whenever possible. The number of patients affected can usually be obtained from national healthcare statistics sources and medical literature. Other good sources are the medical specialty associations such as the American Heart Association, American Medical Association, etc. For example, if the device is intended for use on neonates, the number of neonates treated each year can be found from the American Hospital Association. This will tell you the market for the device, assuming that the device is applicable to all neonates. If, for example, the device only treats neonates with jaundice problems, you need to find the proportion of neonates with this problem and multiply it by the total number of neonates to arrive at the potential market for this device. Documenting your calculations here will lend credibility to your final report.

2.1.3 DOES IT CREATE NEW PROBLEMS?

This is a question that, unfortunately, is frequently left unanswered in final assessment reports. In many cases, the new technology causes the displacement of existing technologies or personnel. Also, it could change the work processes within a hospital department and require retraining of the staff. Overcoming these self-created problems may be a crucial issue for the acceptance of the new technology. Plans and costs to overcome these issues should be included in the report as a part of the process and expense to implement the new technology.

2.2 IS IT CLINICALLY TESTED?

It is vitally important to report the results of any clinical studies[1], if any have been conducted. Having no clinical data to show clinical efficacy immediately ends the assessment process with a negative recommendation. It is relatively easy to search and identify clinical studies that have been published using Pubmed, Google Scholar, or other search engines. When reading a study report begin by finding out who funded the clinical trial, and the type of publication in which the results are reported. Regarding the first point, it is important to note whether the manufacturer of the device or system being tested financed the study in whole or in part. If this funding situation exists, it should be disclosed in the assessment report to allow the proper weight to be given to the data given that some bias may exist in the article because of this commercial funding.

Regarding the second point, publications can vary widely in terms of quality. Basing your report on lower-grade published data is not credible and therefore discouraged. A practical clinical evidence rating (highest value to lowest value) is given as follows:

1. Multi-center Prospective Randomized Clinical Trial (PRCT) – peer reviewed journal (highest rating).

2. Single-center PRCT – peer reviewed journal.

3. Accepted abstract.

4. Submitted abstract.

---

[1]Both terms (studies, tests) are meant to define a formal, clinical assessment of a new technology - real use in real settings by real clinicians and allied healthcare professionals.

5. Vendor sponsored study (unpublished).

6. Vendor sales material (lowest rating).

(Note also that even among peer-reviewed journals, there is a substantial range of quality. This can be assessed this by looking up the journal's "impact factor," which indexes how often articles from that journal are cited by others. Data published in a top-tier journal such as the New England Journal of Medicine should generate high confidence in the findings. There are other lower-tier journals, however, that charge authors a fee to "peer review" the articles, but they will essentially publish any article that is submitted. Just because it's in print does not mean it's true!)

Two sub-questions must be investigated to answer the "Clinically tested?" question: is the test design valid, and are the test results clinically meaningful? In order to answer these two questions, knowledge of the different types of clinical studies, their design and how to interpret the results are necessary.

2.2.1   CATEGORIES OF CLINICAL STUDIES

There are two main categories of clinical studies that are used to validate medical technology: prospective, randomized clinical trials and gold standard comparisons. Each of these categories has variations as well, which will be defined shortly. While this is not intended to be clinical research primer, this book contains information about research that is necessary for complete understanding of the 3Q Method. Table 2.1 lists the two clinical trial types, their target technologies, and an example list of trial outcome variables. Outcome variables are those measurements made during the trial to demonstrate whether the particular technology being tested is efficacious and/or effective.

The terms "efficacy" and "effectiveness" are important to understand as they relate to technology in a clinical trial. The efficacy of a technology refers to how well it works for the particular sample of patients considered in a trial and under the particular constraints created by the research procedures. The types of patients and the experiences they have within a clinical trial might not be very representative of the real world. Effectiveness is defined as the benefit observed outside the highly controlled trial atmosphere. It says that the trial results or conclusions can be extrapolated to the general population of patients. For example, a new device for self-monitoring of glucose might prove beneficial (efficacious) in the trial setting, where nurse researchers contact the patients weekly by phone to monitor adherence and answer any questions. Use of the device in a community clinic setting, however, may prove less beneficial (effective) because patients do not use it consistently without the regular monitoring that was provided during the clinical trial. A company would like to see their new product as both efficacious and effective – which is a powerful statement if it can be made.

Multi-center Prospective Randomized Double Blinded Clinical Trials
This is a well-respected type of clinical trial because it takes into account a broad scope of variables. Each of the terms in this name refers to strategies for minimizing the potential for bias in interpreting

Table 2.1: Clinical test types and their target technologies.

Clinical Trial Types	Target Technologies	Typical Outcome Variables
Prospective randomized clinical trial	Measurement systems Therapy systems Clinical info systems	LOS, mortality, hospital readmission rate, infection rate, number of lab tests ordered, error events
Gold standard comparison	Measurement systems Screening systems	Bias, precision, sensitivity, specificity

the results. "Multi-center" means that more than one medical center participates in the study and the results of all the participating centers are statistically integrated to form the study conclusions. Merging data from multiple locations helps to mitigate the slight differences of how one center practices versus another.

"Prospective" means that the study is designed and then the study data are collected. A "retrospective" study, on the other hand, analyzes data collected from a previous study – a study not necessarily designed to analyze the technology under consideration. Therefore, the validity of a retrospective study can be suspect and caution should be used when including retrospective studies in a 3Q report. Prospective studies are much preferred over retrospective studies for technology assessment.

"Randomized" means that in order to eliminate bias, the patients are randomized into a "control group" and a "test group" (new technology group). Any differences observed between the groups is more likely due to differences in what the groups received rather than pre-existing differences in patient characteristics.

"Double blinded" refers to reducing the bias in a study by blinding the patients and the clinicians regarding which arm of the study the patient is placed. Single blinded refers to blinding only the patients or only the clinicians. In some cases, blinding is not possible, e.g., the device being tested is obvious at the bedside.

Summarizing - the best clinical study for a medical device is a multi-center, double blind, prospective, randomized, clinical trial. This does not mean that a single-center randomized trial cannot be used in a 3Q report. In fact, because of the expense of multi-center studies, most studies in the peer-reviewed literature are single center studies.

There are three main sub-types of randomized clinical trials used for medical technology devices or systems.

1. *Parallel two-arm trial* – one arm as the control group and one arm as the new technology group.

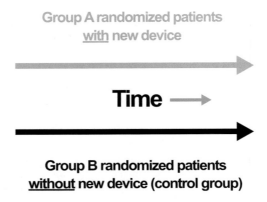

Figure 2.2: Parallel Two-Arm Trial.

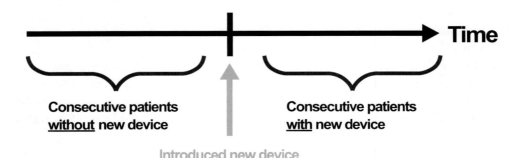

Figure 2.3: Sequential Two-Arm Clinical Trial.

2. *Sequential two-arm trial* – time sequential study whereby given number of patients, usually consecutive, use the existing technology for "n" days and then the new technology is installed in place of the existing technology. Additional staff training on the use of the new technology is usually required. At this point, an equal number of patients, usually consecutive, use the new technology. Data are recorded during the sequential time period regarding patient outcome, cost, etc., and compared at the end of the trial. This type of trial occurs when it is difficult or impossible to have the existing and new technology simultaneously in a care area, (e.g., clinical information system). A disadvantage of the sequential two-arm trial is the possible uncontrollable, changing conditions between the sequential time periods and the fact that two different patient populations are being measured in which there may be different characteristics (age, illness severity, etc.).

3. *Crossover randomized clinical trial* – starts with a parallel two-arm trial and then switches the two arms (crossover) and proceeds with another two-arm trial such that each patient has participated in the control arm and the new technology arm. This method helps reduce any bias introduced by differences in the two patient populations. However, it is subject to changing conditions between the two time periods, costs more, and takes twice the time.

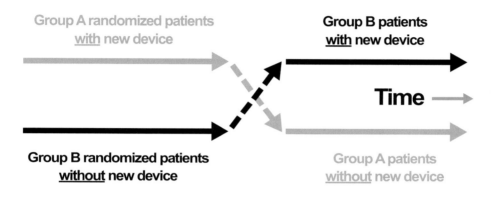

Figure 2.4: Crossover Randomized Clinical Trial.

Statistical Significance

An important factor in the evaluation of a study's results is the statistical p-value of the results. The p-value represents the "statistical significance" of the findings. Technically, the p-value is the *probability* that the results observed in a sample could have occurred if there is no real effect in the population from which our study sample was drawn. For example, in the population of all men in the US, it is pretty likely that left-handed and right-handed men do not differ in average shoe size. It is possible, however, that if we drew a sample of 20 left-handed and 20 right-handed men, the difference between the groups in mean shoe size would not be exactly zero. This could happen simply because of "sampling error." That is, we just happened to draw a sample that did not perfectly resemble the population. Thus, we need a method to help us determine whether a result we observe in the sample is a good representation of what's happening in the population. To reiterate, the p-value tells us how likely we would have observed our sample's results if, in fact, there was no difference (or relationship) in the population. The higher the p-value, the less we can believe that the observed difference in the means is a reliable indicator of the relation between the respective means in the population. For instance, a p-value of .05 indicates that there is a 5% probability of getting a difference in means like the one we observe if there was no different in the population. The p-value is a function of the sample size (> sample size = lower p-value), the distance between the means of the two arms (> distance = lower p-value), and the shape of the two measured populations

($<$ spread = lower p-value). By present medical research convention, the p-values less than .05 are regarded as statistically significant.

Let's look further at how p-values are used. Figure 2.5 (a) and (b) is an example of what the data might look like for a clinical trial to determine if a new ventilator would be a significant improvement compared to the present ventilator (control group). Length of stay (LOS) was chosen as the variable that would tell the investigators if there was an improvement or not. Some liberty has been taken with this drawing below to demonstrate the visual results. Figure 2.5 (a) shows the LOS difference to be 11 days less for the new ventilator with p$<$.05 and (b) shows the LOS to be 1 day improvement with p$>$.05. In the first case (a), since p$<$.05, it means that there is less than a 5% chance that we would observe a difference of 11 days if the real difference in the population was zero. Therefore, by medical research convention, we infer that the new ventilator would reduce LOS in the entire healthcare system from which the study patients were sampled. Another way of stating this case is that p$<$.05 means that the means of the two groups are far enough apart and that the shape of the two patient population curves are narrow enough that they do not intersect with each other very much. Note that the publications usually don't provide the population curves below, but they infer them by reporting the p-value.

Figure 2.5 (b) demonstrates the opposite situation. The mean LOS values of the two arms of the study are only separated by 1 day and the population curves overlap considerably. When the p-value was computed, it was greater than .05 and, therefore, the one-day LOS reduction cannot be applied to the general population of patients.

M.A.A.R.I.E. Method of Clinical Study Analysis

To analyze the validity of a clinical study used for a new technology, the MAARIE Method[2] by Reigleman is convenient and easy-to-use. The MAARIE Method allows one to ask key methodological questions which will aid in identifying the weak areas of a study. This analysis is absolutely necessary if the results of this study will be included on the 3Q report. Failing to validate the study method will greatly compromise the credibility of the final assessment report and, possibly, a purchase decision by a company or hospital. Figure 2.6 describes the MAARIE randomized trial flow – from study design to results interpretation.

Under the "Method" category, these are the items for consideration:

1. Are there valid inclusion and exclusion criterion for patients? For example, if the study purposely left out a certain kind of disease patients, the study results may look good but the potential product could not be marketed to the patients who had the disease population left out of the study. Thus, the commercial opportunity could be greatly diminished and less likely to be funded.

2. Is the sample size sufficient? A qualified biostatistician should make this determination such that the study results are statistically significant. Usually peer-reviewed study publications

[2] Riegelman RK: Studying a study & testing a test (5th edition). Philadelphia, Lippincott Williams & Wilkins, 2005, pages 2–15.

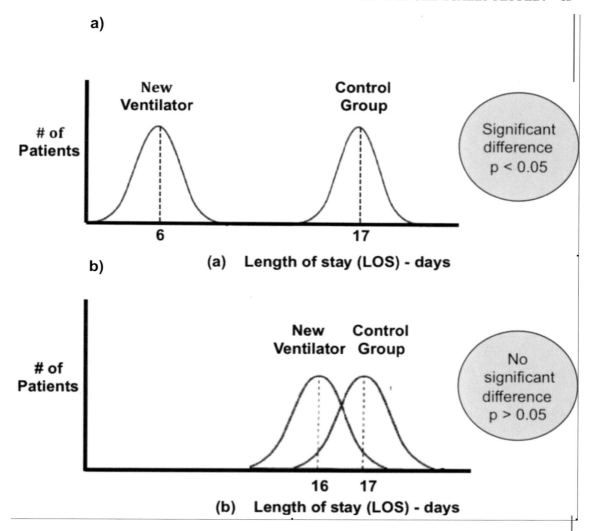

Figure 2.5: Ventilator Example of Clinical Trial Results.

employ professional biostatisticians. Non-published studies done by companies may not have an adequate number of patients to make the results statistically significant. This is certainly an issue that must be identified in the 3Q report.

3. Do the outcome variables appropriately answer the study question? For example, if a new ICU ventilator is being tested, the investigators might want to answer the question, "Does the new

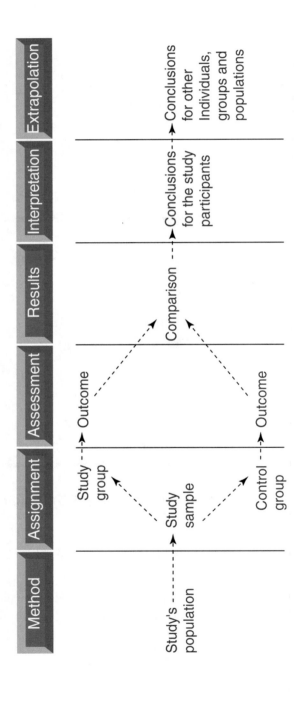

Figure 2.6: M.A.A.R.I.E. Framework for a Clinical Study.

ventilator reduce the incidence of pneumonia?" Some viable outcome variables to measure during the trial might be rate of pneumonia, days on the ventilator and weaning time. This collected data would help pinpoint whether or not the incidence of pneumonia is impacted with the new technology.

Moving to the "Assignment" category in the MAARIE framework, the key questions are:

1. Is there a random patient assignment? If not, this study should, in most cases, be excluded from the assessment process. It is important to have a random sample in a clinical study. The only exception might be if a study included all consecutive patients, admitted to an ICU, for example. In this case, there is no personal selection in who becomes sick enough to end up in the ICU.

2. Are investigators in the trial or study blinded? Study results are compromised when the clinicians who are making the patient decisions are not blinded. For example, if a clinician knows that the patient was assigned to the true device rather than the sham (fake) device, that clinician might unknowingly convey a more optimistic attitude toward the patient, possibly changing the outcomes for that patient. In some cases, it is impossible to blind the clinicians, for example, comparing the patient outcomes when using two different ventilators. If blinding of the clinicians is not possible, it would be prudent to consider that the data collectors in that case should be blinded.

When in the MAARIE method's "Assessment" phase, there are key concerns:

1. Are measurement method(s) used to measure the outcome variables accurate enough for this study? For example, take the study question is, "Is the number of blood gases ordered per day decreased by the use of a new technology?" If the data collection protocol states that patient blood gases should be recorded only during the day shift, then the measurement method is insufficient to meet the objectives of the study since the measurements are not made over 24 hours.

2. Is there any possible bias during the study on the part of the clinicians, patients, or data recorders? At the assessment point of the study, any possible biases should be investigated.

Under "Results," there are key questions:

1. Is there a significant difference between two groups ($p < .05$)? Any p-values greater than .05 are enough to exclude the study from this technology assessment, assuming the study had an adequate sample size.

2. Have any results, (patients) been thrown out to make the statistical results look better?

Within the MAARIE Method, there are key "Interpretation" questions:

1. Is there valid clinical interpretation of the statistical results? Do the study authors correctly apply the results to the original study question?

2. Are the results clinically significant? Even though the study results are *statistically* significant ($p < .05$), the new technology may or may not be useful in actual clinical settings. The "efficacy" of the technology must be assessed and detailed in the 3Q report. To answer this question, the 3Q investigator must determine how the new or improved measurement, therapy or information will benefit the clinicians and patients, and to what extent. Is it a compelling benefit or a small benefit? For example, a new ventilator that decreases the incidence of pneumonia by 1% may not be worth purchasing because the incremental cost is much higher than the expected benefit. In summary, the new technology must pass two tests: a) statistical significance, and b) clinical benefit.

3. Have all adverse effects or events been identified? Almost every new technology generates new problems. Anticipate what those may be and report them.

Finally, under "Extrapolation" there are key questions:

1. Are the study results so compelling ($p < .05$) that the results can be reasonably expected in the general population of patients? Studies that include more than one institution often are more compelling than single institution results because they cover a multitude of clinicians, care protocols, and patients. Most new technology is tested in multiple institutions before they are released as products. So if one is evaluating a new technology which has only one study and $p < .05$, the most that can be concluded is that this technology is worthy of continuing follow-up.

2. Can it be projected what the real benefit will be outside the somewhat artificial trial atmosphere? In short, the effectiveness of the technology must be measured to determine to what extent the trial results or conclusions can be extrapolated to the general population of patients.

Gold Standard Comparison Studies

There are typically two types of Gold Standard[3] assessment studies for medical technology:

1. Measurement devices, (e.g., blood pressure).

2. Screening systems, (e.g., mammography).

Gold Standard Comparison Study – Measurement Devices

Measurement device studies determine the accuracy of a new measurement technology compared to the existing clinical "gold standard" technology (e.g., a new bedside blood gas instrument compared to the existing hospital clinical lab instrument). The basis for the quantitative comparison of a gold

---

[3]In medical terminology, a measure of comparison against that which is considered ultimate or ideal; the test of the value of a procedure that is considered definitive.

standard device and a new device is the Bland Altman[4] analysis. This analysis provides a method for deriving meaningful statistics and plotting the results that allow you to make a value judgment of how closely the new method approximates the accuracy of the existing gold standard method. Note that it *will not* indicate if the new method is *more* accurate than the present gold standard method. This assumes that the present gold standard is not a direct measurement, but it is the present best method of measurement. In this hypothetical case, the inventor or company might believe their new device will be much cheaper to produce and thus take significant market share away from existing devices, but it has to perform up to the gold standard or the users will reject it.

Let's look at a typical Bland Altman example. Table 2.2 contains data for an existing blood pressure device (gold standard) and a new blood pressure device. The first column shows that eight different comparison readings were recorded for this subject. The next two columns show the values recorded for both the gold standard device and the new device. Columns four and five calculate the two variables of the Bland Altman analysis: the mean of the gold standard device and the new device for every pair of points, and the difference between the new device and the gold standard device for every pair of points. These two data sets are then plotted against each other, creating the Bland Altman plot.

Table 2.2: Example of raw Bland Altman data.				
Patient #1 Systolic Blood Pressure Test			**Bland Altman Plot data**	
	Gold Standard		**Mean**	**Difference**
	Device	**New Device**	**(mm of Hg)**	**(mm of Hg)**
Reading	**(mm of Hg)**	**(mm of Hg)**	**(x axis)**	**(y axis)**
1	122	120	121	-2
2	116	112	114	-4
3	110	102	106	-8
4	97	89	93	-8
5	111	101	106	-10
6	134	127	130.5	-7
7	145	151	148	6
8	150	159	154.5	9
9	144	135	139.5	-9
10	110	102	106	-8

Below is the corresponding Bland Altman plot for the data in Table 2.2 above. This data represents the data from Patient #1. In a real situation, data from a large number of subjects (50 or greater) would be included in the Bland Altman plot.

---

[4]Bland JM, Altman DG: "Statistical methods for assessing agreement between two methods of clinical measurements." The Lancet February 8, 1986, pages 307–310.

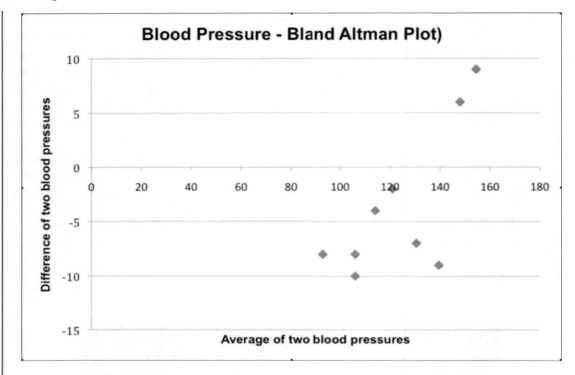

Figure 2.7: Bland Altman Plot for the Patient in Table 2.2.

The Bland Altman plot is a key element of the 3Q Method when the technology is used for some method of measurement. The plot provides valuable information that will ultimately help answer the "Is it real?" question. It provides three quantitative statistics: bias, precision and limits of agreement. These three statistics are widely accepted as a measure of how well a new technology compares to a gold standard technology.

1. *Bias* – the systemic tendency to deviate from the true value of points. Bias is the mean difference between the gold standard points and the new device points.

2. *Precision* – the scattered deviation from the true value of all test points as measured by the standard deviation. Precision is the standard deviation of the differences between the gold standard points and the new device points.

3. *Limits of agreement* – ±2 standard deviations (SD) that describe the range of 95% of the comparison points, given that the distribution of data points is roughly a normal Gaussian

distribution[5]. This "2 SD range" of a clinical variable means that the clinician can expect that 95% of the measurements made with this new device will be between ±2 SD. For example, a comparison study for a new blood pressure device may yield a SD of 5 mm of Hg. The limits of agreement are ±2 SD or ±10 mm of Hg in this case. The 3Q investigator must then find out if the clinicians can live with 95% of the measurements lying between ±10 mm of Hg of the actual value. For systolic blood pressure, this would mean if the actual value is 120 mm of Hg, 95% of the time the new device would read ±10 mm of Hg, or between 110 and 130 mm of Hg. The clinician then must decide if this error is tolerable in the clinical setting. If it is not, then the new device "fails" the precision test and probably fails the 3Q analysis.

In general, if the present clinical acceptability range is larger than the Bland Altman 2 SD range, the new technology is acceptable from a precision perspective, assuming there are an adequate number of measurement points across the clinical range of the variable.

How do you decide what bias and precision for a given physiological variable are clinically acceptable? As a good rule of thumb, an acceptable bias value for virtually any measurement between the means of the new device and the gold standard would not be greater than 3-4%.

Arriving at a clinically acceptable precision value is not an easy task since the clinicians are not used to considering this statistic. One way to determine if the precision is acceptable is to look at the clinical and biomedical literature dealing with similar technology. If these are peer-reviewed journal articles and they conclude that a new technology is acceptable for clinical use regarding bias and precision, then this will provide a set of acceptable bias and precision values.

Another source for bias and precision values is the FDA or other groups that publish acceptable accuracy values in journals such as the American Association of Medical Instrumentation (AAMI) or the American Association of Anesthesiologists (ASA). Keep in mind that if any journal article is five years old or older, these bias and precision values may not be applicable today. Using multiple, up-to-date publications as sources will give the 3Q report additional credibility.

Still another way of acquiring acceptable precision values is to ask expert clinicians directly, "How much of a change in the physiological variable must there be before you change your therapy regime?" For example, if a clinician responds, "If the blood pressure changes by 15 mm of Hg, I will alter my medications," one can conclude that changes of ±15 mm of Hg are clinically significant. It could then be assumed that the measurement system should be at least ±0.5 of that (or ±7.5 mm of Hg), to cause the clinician to start or change therapy with an otherwise stable patient. So if 95% of the measurements lie within ±7.5 mm of Hg, which provides a desired precision goal. This last method is one that can be used when other sources fail to provide good clinical estimates. Asking three to four clinicians this question would be necessary to make sure the precision estimate is representative.

These three statistics are graphically represented in Figure 2.9 below. A good clinical publication will illustrate the Bland Altman plot with statistics plotted and numerically annotated.

[5]In probability theory and statistics, the normal distribution or Gaussian distribution is a continuous probability distribution that describes data that cluster around the mean.

Some publications will also show a correlation coefficient plot with the correlation coefficient, (e.g., $r = 0.94$). This correlation coefficient is useful, but it should always be accompanied by the Bland Altman plot. The reason is that r conveys only some of the information about how two sets of values agree. To make this clear, imagine that the gold standard values in a dataset are 10, 20, 30, 40, and 50, whereas the new device's values are 60, 70, 80, 90, and 100. The correlation between the two measure is $r = 1.00$, the highest possible correlation. But it should be clear now that the Bland Altman plot would make the new device look terribly biased.

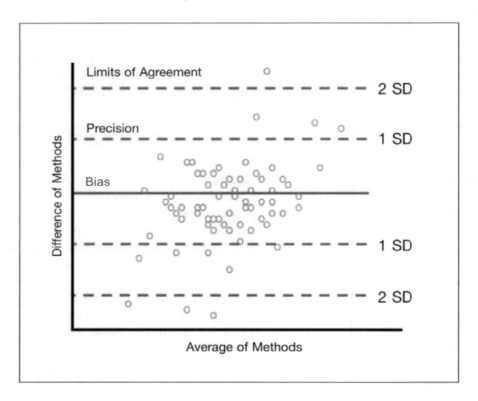

Figure 2.8: Bland Altman Plot Showing Bias, Precision, and Limits of Agreement.

Besides providing the three important statistics – bias, precision and limits of agreement – the Bland Altman analysis provides a visual plot of the *mean* of the two measurement systems (x-axis) versus the *differences* of the two measurements (y-axis). This plot shows if there are sufficient test points along the entire range of the measured variable. It is necessary to have data points in sufficient quantity across the entire clinical range of this variable. Sick patients' values usually occur at either ends of the x-axis, so the plot must contain data points at the extremes, not just in the middle. This is easy for the eye to see, which is another benefit of the Bland Altman plot.

For example, in measuring adult blood pressure there should be data points over the range of sick adult patients, e.g., from 40 mm of Hg to 180 mm of Hg. If there were not, even though the bias and precision may be clinically acceptable, this study would have to be rejected for lack of extreme data. Figure 2.9 is an example of a Bland Altman plot of blood pressure with too few data points at low blood pressure. Ironically, it is sick patients with low blood pressure who would stand to benefit from this proposed technology

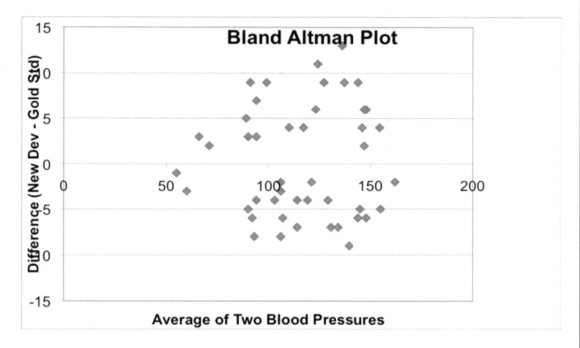

Figure 2.9: Bland Altman Plot with Too Few Data Points.

Bland Altman results should be reported in their entirety. Figure 2.10 illustrates the preferred slide format for presenting the Bland Altman results. The statistics are shown in the upper right, i.e., n (number of data points), bias and precision. The Bland Altman plot is shown just below the statistics with the bias, precision and limits of agreement delineated so the reader can quickly scan the plot. The numbered list on the left draws the clinical conclusions from the statistics and the plot.

Gold Standard Comparison Studies — Screening Systems
Screening systems measure certain physiological or anatomical variables that determine if a person does or doesn't have a specific disease. Some screening measurements simply detect the presence of the particular disease such as cancer screening. Some systems actually measure a variable and test if it

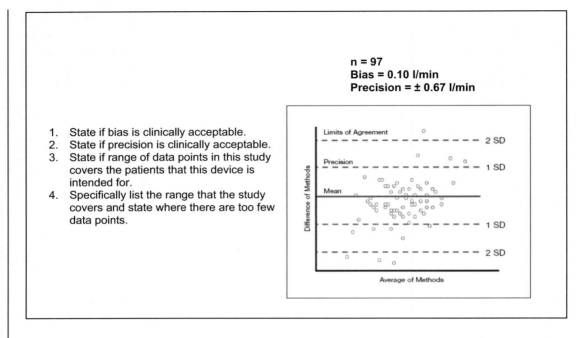

Figure 2.10: Slide Layout for Reporting Bland Altman Results.

is below or above a certain "cut point" value. For example, a blood pressure greater than 140/90 mm of HG would be the "cut point" value of hypertension. Cut point values are usually defined by national committees of expert clinicians. New screening technology is continually being developed that reduces the cost of existing screening systems, or introduces a new variable that may be a better precursor to disease detection. These new technologies need to be evaluated against existing clinical gold standard screening systems.

All screening systems are either based upon a single measurement cut point or a set of diagnostic rules. The key challenge in designing these systems is to place the cut point or create rules such that the technology or system produces the optimum clinical accuracy. Few of these systems are 100% accurate because of disease complexity. Of special concern is the cost and patient impact of inaccuracies in a screening system. False results may lead to further testing, additional cost, patient inconvenience, or even morbidity and mortality. The 3Q process considers the nature of these screening system inaccuracies using the concept of true positives and negatives and false positives and negatives.

The point of the 3Q assessment is to determine the performance of a new technology compared to the gold standard measurement, given the cut point or rules inherent in the system. The results of these comparison studies are usually expressed in terms of true or false readings compared to the gold standard. There are four possible accuracy "states:"

- *True positive (T+)* – test system and gold standard system agree person has the disease.

- *True negative (T−)* - test system and gold standard system agree person does not have the disease.

- *False positive (F+)* – test system says person has the disease and the gold standard system indicates person does not have the disease.

- *False negative (F−)* – test system says person does not have the disease and the gold standard system indicates person has the disease.

These possible conditions can better be visualized using a Truth Table.

Table 2.3: Truth table for assessing screening tests.

		Gold Standard (Truth)	
		Positive	Negative
Test Results	Positive	True Positives	False Positives
	Negative	False Negatives	True Negatives

By convention, the results of these comparison studies are usually presented using four statistics:

- *Sensitivity* = true positives/(true positives + false negatives).

- *Specificity* = true negatives/(false positives + true negatives).

  Positive Predictive Value (PPV) = true positives/(true positives + false positives).

  Negative Predictive Value (NPV) = true negatives/(true negatives + false negatives).

Usually in a given screening system there is an inverse relationship between false positives and false negatives. In other words, as the cut point or rules are adjusted to reduce false positives, the false negatives increase. Determining what level of false positives and false negatives are clinically acceptable for a specific screening system is difficult and depends on the assessor seeking clinician agreement.

An example that illustrates the resultant variability of false positives' (F+) and false negatives' (F−) by adjusting the detection algorithm is the case of automatic arrhythmia detection systems used in almost all EKG patient monitors today. The main purpose of these detection algorithms is to detect premature ventricular contractions (PVCs). PVCs can be a harbinger to more serious arrhythmias that can cause patient death. Both F+s and F−s are of concern to the clinicians who care for these arrhythmia patients. A system with low F−s is beneficial, because high F−s might result in no medical intervention when it is actually needed. On the other hand, excessive F+s cause the clinical staff to respond when no response is necessary, possibly endangering the patient with unneeded therapies and wasting staff time. It is well known that clinicians tend to turn off the arrhythmia alarms if the F+ rate is "too" high, which could possibly compromise patient safety. Therefore, it falls on the algorithm designers to minimize both F− and F+s in healthcare technology.

Arrhythmia algorithms usually consist of making several ECG waveform measurements and creating rules based upon two beat-to-beat variables: width and interval. Since variation in these two ECG measurements exist in the normal patient population, the arrhythmia algorithms have to take into consideration what width and interval actually define a PVC. For instance, a set of sample tests run on an arrhythmia algorithm might look like Table 2.4 below. Usually these algorithms are tested against a set of standard ECG tapes in which each beat has been classified by a cardiologist(s) and considered to be the "gold standard."

Table 2.4: Example of arrhythmia algorithm performance.				
Algorithm version	PVC width (ms)	PVC R-R Interval (ms)	F+ (%)	F− (%)
Version 1	> 80 ms	< 700 ms	12%	3%
Version 2	> 90 ms	< 500 ms	5%	10%
Version 3	> 100 ms	< 400 ms	2%	27%

In the above example, various values of F− and F+s can be achieved by altering the definition of PVC width and interval. It is the company's responsibility to choose which combination of these two algorithm variables to market. In the arrhythmia detection case, the company might pick Version 2, which accepts a F+ of 5% and a F− of 10%. This example has been purposely simplified to demonstrate the principles involved with truth tables for assessing screening tests.

Now that both randomized clinical trials and gold standard comparisons have been described, a summary table of this important information is given in Table 2.5.

Table 2.5: Summary of clinical testing methods.

Types of Clinical Testing	When Used?	Key Analysis Tools
1) Randomized Clinical Trials (RCT)	Testing clinical effect of a new technology via two arms	M.A.A.R.I.E.
2) Gold Standard Comparison – measurement device	Comparing accuracy of new measurement device with gold standard	Bland Altman Analysis M.A.A.R.I.E.
3) Gold Standard Comparison – screening test	Comparing accuracy of diagnostic screening test with gold standard	T−, T+, F−, F+ Sensitivity, Specificity M.A.A.R.I.E

2.2.2   CLINICAL LIMITS TESTED?

In order for new medical technology to be an effective tool in the healthcare delivery system, the technology must be tested over a wide *range of patients* and *patient conditions* in actual clinical use. Technology tested over a limited range of patients and patient conditions usually will not survive the rigors of clinical medicine. Products that reside in hospitals and clinics will be used by a wide variety of patients with a wide variety of conditions, so it is necessary to question the clinical limit conditions under which the technology has been tested. The following is a list of 3Q questions that target these clinical limits. The task of the 3Q assessor is to understand the variety of patients and clinical situations to which the technology would be applied and to determine if the new technology has been tested under these conditions. If the new device has not been tested under these conditions, then this fact compromises the use of this technology and must be noted in the final 3Q report.

Has this new technology been tested…

- over pertinent patient age range?

- over pertinent patient weight range?

- over wide physiological variable range?

- with types of drugs commonly used?

- with types of non-drug therapies used?

- with various common patient co-morbidities?

The clinical limit requirements are obtained by defining the *range of patients* and the *range of patient conditions* to be seen by the specific technology. For example, a new blood pressure monitor intended for use in the ICU might have the following clinical limits:

| Table 2.6: Clinical range for new blood pressure monitor example. ||
Clinical Item	Clinical Range
1) Patient age	18 – 100 years
2) Patient weight	70 – 300 lbs
3) Blood pressure	60/40 – 225/180
4) Drugs used	Vasoactive, chronotropic, ionotropic
5) Other therapies	Ventilators, IV pumps
6) Patient co-morbidities	Diabetes, congestive heart failure, pulmonary hypertension, aortic valve regurgitation & stenosis, septic shock

If the clinical study does not cover the ranges in Table 2.6, then the omissions must be reported in the 3Q report. These omissions may have significant repercussions for the applicability of the new blood pressure technology. For example, if the device was only tested for the blood pressure range of 90/70 to 225/180 mm of Hg, then it would have questionable accuracy for sick patients below 90/70 mm of Hg, and clinical decision errors could occur. If it was not tested on patients who were on vasoactive meds, then the accuracy on these patients cannot be defined.

Note that the above list of patient range items is not an exhaustive list, and other more appropriate items for other technologies certainly exist. It is left up to the assessor to determine the key range items that must be considered. However, these six items are a good starting point.

Where does the 3Q assessor get the clinical information contained in Table 2.6? Here are several sources:

- Specifications of existing products.

- Specifications published by various medical societies, e.g., American Society of Anesthesiologist (ASA).

- Discussions with clinical experts.

- Clinical visits.

- Specifications published by standard setting groups such as American Association of Medical Instrumentation (AAMI), Health Level 7 (HL7).

2.2.3   CLINICAL USABILITY

For effective use over time, a technology must enhance clinician workflow. Many product benefits are associated with their usability. Some of the main features to investigate are given below. Any aspects of the new technology that detract from the intended clinical use must be reported in the final

3Q report. This information is best obtained by visiting the clinical area where the new technology will be used and seeing the technology in use, or beforehand to screen for these optimal requirements.

1. Appropriate size and weight?

2. Mobility?

3. Display size; visualization?

4. Control location; ease of use?

5. Cleaning; sterilization?

6. Safety risk to patients and workers?

2.2.4   TRAINING COMPLEXITY?

The amount and complexity of technology training can make or break a medical device or system because of the cost to the hospital. These costs consist of any outright training costs, as well as the harder-to-document learning curve that impacts productivity and revenue. This 3Q area must take into account all personnel that must be trained on this technology – clinicians, technicians and administrative workers. The amount and cost of training will differ depending on the class of users, e.g., doctors vs. nurses, the size of the staff to be trained and the staff turnover rate. The manufacturer and the hospital will usually share training expenses, so both parties are typically interested in this part of the 3Q assessment. Training information can be obtained from manufacturers and institutions already using the new technology. A good method is to look at the training requirements of similar, existing technologies in the hospital or clinic.

1. Training time and method?

2. Typical training cost for the hospital?

3. What percent of training costs absorbed by the vendor?

4. Quality of manufacturer's training materials?

5. Anticipated learning curve for optimal product performance?

2.3   IS IT BASED UPON REAL SCIENTIFIC PRINCIPLES?

Most technologies work because they are based on known scientific principles. Thus, your confidence that the technology is real is increased when there is a recognized scientific principle(s) behind the technology. Some technology has a long history, (for example, x-ray machines), and no consideration of this question is necessary. However, if the technology has little history or data, the scientific principles and the extent of applicable scientific literature must be researched. Information systems are usually not based upon physics or chemical principles and need no analysis under this question. Some technologies have little scientific basis as we know it today, e.g., empirically derived algorithms. In these cases, the main investigation will center on the clinical testing of the technology to make sure that it achieves the specified clinical purpose.

CHAPTER 3

Question #2: Can We Win?

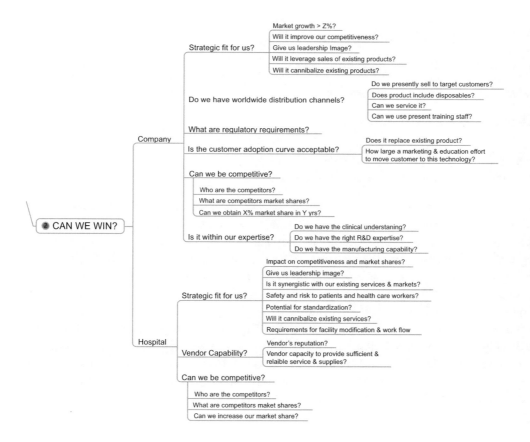

Figure 3.1: 3Q Method Map – Question #2: Can We Win?

This section answers the question, "Can our hospital or company acquire this technology and win…given our strengths and weaknesses and the competitions' strengths and weaknesses?" Since there are several areas under this section that differ depending on whether the analysis is from a company or hospital perspective, each of these will be explored separately.

3.1   CAN WE WIN (COMPANY)?

Under this major 3Q question, six separate areas will be addressed:

1. Is it a strategic fit for us?

2. Is it within our expertise?

3. Can we be competitive?

4. Do we have worldwide distribution channels?

5. Is the customer adaptation curve acceptable?

6. What are the regulatory requirements?

   These six areas will be covered below.

3.1.1   IS IT A STRATEGIC FIT FOR US?

Four areas must be examined to answer this strategic fit question:

A. Most companies will not invest in a new technology targeted for a market that has a minimal sales growth rate per year (e.g., 1-5%). It is therefore important to define the growth of the target market in quantitative terms and compare it with the company's minimum market growth[1] requirement using the equation: Market (sales) growth > Z% with Z as the company's minimum growth requirement for markets using the new technology). The company's strategic plan and target market (domestic versus worldwide) impacts the minimum growth requirements as well, and should be included in the 3Q report. If the technology under assessment is targeted for a slow growing market, then this factor alone could halt the assessment with a negative recommendation. In a brand new market, potential growth may be a challenge to estimate since there is no market history to consider. In this case, it is best to look at other new technologies in similar markets to estimate the market growth rate. The rationale for this estimation must be clearly documented in the final report.

B. Consideration should be given if the new technology will give the company a competitive advantage. This could be as the leader in their industry to be first-to-market, or as a strategic piece of their product line. Specific competitive advantages should be researched, qualified and presented within the 3Q report.

C. In today's market, having a strong leadership image can mean significantly higher sales at higher margins. Maintaining this image requires constant effort, but very essential in the dynamic healthcare industry. The new technology under study should contribute to, or improve

---

[1]"Market growth" is defined as the yearly percentage change in all sales ($) by all competitors.

Table 3.1: Market growth calculation example.

Company	2007 Sales ($M)	2008 Sales ($M)
Company A	24.5	25.1
Company B	17.3	20.3
Company C	14.1	16.1
Company D	5.7	6.2
Company E	3.2	3.7
Total sales	64.8	71.4

2007 – 2008 Market Growth Rate = 10.2%

the company's leadership image in a significant way. Companies typically have a mission statement that infers the specific market image they desire. Mission statements might include: "innovator," "technology leader," "quality leader," "customer service leader," or "low price leader." If the new technology easily correlates to the company's image, acceptance will be easier to establish.

D. Companies with multiple product lines are often interested in whether the new technology will enhance the sales of their existing products. Obviously, a new technology with the ability to "pull-through"[2] sales of existing products is more attractive to a company. It would be appropriate to quantitatively estimate the "pull through" sales that would be realized and document this in the 3Q report. This will be covered further under "Is it worth it?"

3.1.2 IS IT WITHIN OUR EXPERTISE?

Do we have the right research and development expertise? Integrating new technology into a company's product line(s) usually requires engineering resources. If a company is missing the correct engineering expertise (e.g., a software company integrating a chemical technology product), then the report should highlight this deficiency. Ramping up R & D requires significant expenditures of time and money, which must be researched, quantified and documented.

Do we have the manufacturing capability? Some new technology may require a manufacturing process or expertise that differs from what is currently in place, (e.g., printed circuit boards vs. chemical sensors, or software vs. hardware). The report should highlight any differences so that the company is not blind-sided with possibly expensive manufacturing start-up costs.

Will it cannibalize existing products? New technology ultimately obsoletes existing products. Given this reality, the desired obsolescence timeframe should be detailed, with a plan in place to move customers from old to new as seamlessly as possible. Of course, the new product sales should

---

[2]Pull through" refers to a new product that will cause a customer to not only purchase the new product but to purchase some of the company's existing products.

yield higher revenues than the cannibalized current product sales. More on this topic is found in the section, "*Is it worth it?*"

3.1.3   CAN WE BE COMPETITIVE?

Who are the competitors? If the company is planning to enter a new market with the new technology, it's mandatory to be familiar with those who control most of that market. Knowing the main competitors' strengths and weaknesses is key for a high quality, comprehensive analysis. Breaking into a market dominated by one or two companies requires quite different strategies and expenses than entering a developing market.

What are competitors' market shares? In anticipation of entering a new market, the present competitors and their percentage share of last year's product sales should be analyzed and reported. For instance, if there are two or three established companies offering the same technology with a total of 75% market share, a management team might think twice about trying to enter that market. On the other hand, if there are ten companies in this market, each with 10% market share, a management team may see a market penetration opportunity with a unique new technology.

Can we obtain X% market share in Y years? Answering this question with supporting documentation will greatly aid management's decision to pursue new technology. Companies usually include market share goals for new products in their strategic plans, and reporting potential new technology against these goals is mandatory. This requires a prediction of attainable market share by year, over the next five years. *Note:* This is not to be confused with strategic fit, which deals with market growth. Market share deals with what percentage of the total market sales in any given year a specific company achieves. Both market growth and market share metrics are of great business importance.

3.1.4   DO WE HAVE WORLDWIDE DISTRIBUTION CHANNELS?

Do we presently sell to all target customers? This question allows reporting of a situation in which the study technology necessitates selling to a new customer, (e.g., present company products are aimed at cardiologists, and the new technology is aimed at radiologists). Understanding needs and applications of new customers is key to market success of a new product, and it should be investigated to see if expanding or changing the distribution channel is merited. Establishing new distribution channels in today's global market is a costly and lengthy process, and it must be approached cautiously and with sufficient planning. This can be accomplished by growing a company's own sales force or partnering with an established firm that has an existing sales force in those markets. The purpose of the 3Q report is not to solve the distribution issue but to document this challenge.

Does the new product include disposables? This is an extremely important business question. In some cases, sales revenue of disposables can be greater than the main product. If the technology being studied includes disposable items that are used for each patient, (e.g., ECG electrodes), this point deserves special treatment in the 3Q process. Company management will be very interested in the nature of the disposable(s) – the estimated cost and estimated selling price. Profit margins for disposables are typically significantly higher than for hardware devices and, therefore, very attractive. However, selling disposables can be a very different process than selling the devices that use the disposables. Many times the target customer for disposables is the hospital purchasing agent, not the clinician who makes the hardware decision. Acquisition methods and pricing of disposables versus hardware vary greatly as well. These important factors lead some companies to employ a separate sales force that does nothing but disposable sales. Disposable content of any new technology and its inherent differences must be addressed in the 3Q report.

Can we service it? This question is similar to the distribution channel question. If the existing field service team cannot service the new technology without additional training or staff, it should be reported in the 3Q report. Service teams are extremely expensive to develop, train and maintain. In addition, different medical products can have very different service response requirements, depending on their application and where they are used in the hospital. It is important to compare the required response time for the new technology with the company's present service force response time. Providing customers with a quicker response time can be a significant increase in cost, and it should be well documented in the 3Q report.

Can we use the present training staff? New technology necessitates that the customer be trained on its proper clinical use. Depending on the product or technology, the training could be minimal or extensive, with an associated scale of costs. There is usually some training included with the initial product placement, but additional training costs should be researched and documented. Hidden behind training costs may be replacement staff expenses while individuals are training. This leads to the questions such as: "Must four lab technicians or 140 nurses be trained?" For a simple product installation, the sales force might do the training around employees' schedules with minimal downtime impact to the facility. More complex product training might involve a special company training team at the customer site for a week, (e.g., training 100 nurses for 30 beds of patient monitoring equipment). It is critical to ask this question and detail all aspects of the training requirements.

3.1.5   IS THE CUSTOMER ADOPTION CURVE ACCEPTABLE?

How large a marketing and education effort is required *pre-sale* to move the customer to buy this new technology and its benefits? This differs from the post-sale education on how to use the specific product or technology covered previously. For example, a new technology that introduces the measurement of a new physiological parameter requires customer education about the benefits and patient impact of this technology. The company would most likely devise a market campaign to stress the benefits of incorporating this new technology in their practice. These educational market campaigns are usually very expensive, and, therefore, this extra cost must be assessed and highlighted in the 3Q report.

Does it replace an existing customer product? If the new technology replaces an older, existing customer system or device (e.g., old X-Ray system being replaced by newer model), then the customer should already be familiar with the application of the technology. In this case, the customers will need little instruction and sales of the new technology will ramp up relatively fast. However, if the new technology is not familiar to the customers, adoption and learning curves may be lengthened, impacting the sales cycle and rate of new sales. Customers' willingness to embrace a new technology is extremely important to company management. Given everything being equal about two new technologies except the customer rate of acceptance, the company is likely to pick the shorter adoption curve to maximize revenue from their investment.

3.1.6 WHAT ARE THE REGULATORY REQUIREMENTS?

There are two important 3Q report issues regarding regulatory requirements: a) Has FDA approval been obtained, and b) What FDA Class is this new technology? Both the approval and FDA Class documentation should be included in the 3Q report. There are three FDA Classes (I, II, III) with appreciably different costs to the company associated with each. Class I devices present minimal potential for harm to the user and are often simpler in design than Class II or Class III devices. Class II devices are those for which general controls alone are insufficient to assure safety and effectiveness, and additional existing methods are available to provide such assurances. Therefore, Class II devices are also subject to special controls in addition to the general controls of Class I devices. Special controls may include special labeling requirements, mandatory performance standards, and post market surveillance. A Class III device is one for which insufficient information exists to assure safety and effectiveness solely through the general or special controls sufficient for Class I or Class II devices. Such a device needs premarket approval, a scientific review to ensure the device's safety and effectiveness. Determining the FDA Class for the new technology will aid in determining some of its associated costs.

Company patents describing the new technologies are vitally important in today's business world and another consideration for the 3Q Method analysis. In most cases, worldwide patent filings are more valuable than strictly domestic filings because of the potential worldwide distribution of the technologies. The absence of patents would signify that the new technology is still in a very early stage. A well-written patent can serve as a barrier to a competitor's entry into these technology areas. Any patent information filed on the new technology should be highlighted in the report.

3.2 CAN WE WIN (HOSPITAL)?

In today's healthcare market, hospitals are usually organized into geographical delivery systems with multiple hospitals, clinics, etc., managed by a large enterprise. This enterprise has a strategic plan that helps guide the acquisition of new technology at all points of care. The "Can we win?" question considers the hospital's adoption of any new technology consistent with this strategic plan. The areas of consideration vary slightly from the company scenario. They are as follows:

1. Is it a strategic fit for us?

2. What is the vendor capability?

3. Can we be competitive?

These three questions will be covered below.

3.2.1   IS IT A STRATEGIC FIT FOR US?

Many areas must be examined to answer this strategic fit question:

Does the new technology provide a competitive edge and/or increase market share? If the technology allows the healthcare facility to enter a new geographic market, new patient segment or new service line, the impact on the enterprise's total patient market share should be estimated. The potential increase in market share will be what the hospital gains by investing in this new technology, and the hospital management will require this estimate. In other cases, the hospital may not gain market share, but they may loose market share if they *don't* acquire the technology, i.e., competitive hospitals already have the technology. In some cases, the purpose of acquiring new technology is not to increase top line revenue but to reduce hospital costs. Both of these factors are motivating reasons for purchasing new technology.

Will it give us leadership image? The hospital's image, both internally and externally, can be impacted by new technology. Some technologies that have high visibility might enhance the public image of the hospital, such as robotic surgery systems, while providing clinicians a leading edge work environment. Thus, the hospital's image is an important factor in technology evaluation. The 3Q report should examine and highlight any potential image gains in these two areas.

Is it synergistic with our existing services and markets? If the new technology is not within the hospital's present services, then the hospital may have to spend extra resources to enter a new service area. The technology assessment must note the possibility of extra resources needed and in which areas: facility, operations, staff, advertising, etc. This could amount to considerable expense for the hospital, in addition to the purchase of the new technology.

Is there safety risk to patients and health care workers? Patient and provider safety are critical elements in today's healthcare system. Any potential risk or safety concern associated with the new technology must be included in the 3Q analysis. New safety procedures, physical modifications to the facility, or certification training might be required, causing potentially significant expense. Many new technologies are bound by federal regulations with strict compliance issues as well. Any of these factors could cause extra expense for the hospital and should be quantified in the assessment report. Representing these and other costs in the 3Q report will be covered in the next section, "Is It Worth It?"

Is there potential for standardization? Having multiple vendors for similar medical technology (for example, patient monitors) is problematic for hospitals due to the added expense of different clinician training and different equipment service. Consequently, hospitals try to standardize on the same vendor in many instances. The question needs to be asked if the new technology could function adequately in all the hospital areas where there is a need for the technology. For example, could an improved pulse oximeter be used in all care areas of a hospital? If so, this is a plus for the new technology and should be noted. This factor is especially important for technology that includes disposables. Volume discounts for disposable products play a significant role in hospital purchases, and, therefore, hospitals like to standardize disposables hospital-wide.

Will it cannibalize existing service? There are cases where the acquisition of a new technology will cause an existing technology or service to become obsolete. This factor must be noted and the net financial effect of the technology "switch" must be assessed in terms of revenue and expense for the hospital. This will be important when the five-year financial projection is generated in the next section, "Is It Worth It?"

Is there an impact on facility and/or workflow? The new technology will require significant hospital expense if the physical plant must be modified to accommodate the new system. This modification could be costly in a number of ways; it may even shut down revenue-generating services during construction. Clinical or administrative workflow may also be impacted by the use of the new technology. Optimal integration of a new product or system requires training for all impacted personnel. Calculating the number of personnel to be trained, the training hours, the pay rate of the personnel taken out of clinical service during training, and the cost of the training staff and materials can approximate the cost of this training. These will be important cost estimates to consider in the five-year financial projection, generated in the next section, "Is It Worth It?"

3.2.2   WHAT IS THE VENDOR CAPABILITY?

What is the vendor's reputation? Since the hospital will usually be entering into an extended business relationship with the vendor of the new technology, it is important that the vendor have a history of good business practices and financial viability. Years in business, past history of customer relationships, stock market performance, management strength, etc., should all be considered in the company analysis. Small, new companies must be thoroughly researched, and significant risks associated with these characteristics must be documented in the 3Q report.

Is there vendor capacity to provide sufficient and reliable service and supplies? Service response, both timeliness and availability of parts, is extremely vital for medical equipment – especially for new products. Sometimes, a manufacturer's service capability will lag during the first shipments of a new product. It can be problematic for a hospital if their newly delivered product has excessive downtime due to parts unavailability or untrained service people. Significant risks in this area should be researched and noted in the 3Q report. This information can come from a variety of sources. Clearly, the vendor can provide their service capabilities regarding response time, availability of parts and service staff training. However, the perspective of hospitals that have dealt with this vendor should also be obtained. It is also useful to ask the vendor for references to contact and interview. One might expect these customers to be the vendors' best customers, but there are usually clinical engineers at these sites who will give objective opinions.

3.2.3   CAN WE BE COMPETITIVE?

Who is the competition? Listing your competitors is especially important if the new technology is adding a new service to the hospital. Other hospitals in the same geographic area might be competing in this same service space but with different technology. Knowing the competition and their practice is an important factor in determining success with a new technology. It can help define a hospital's approach to timing, scope, placement, advertising, etc. Keep in mind that tele-medicine can quickly change the geographic boundaries of traditional markets.

What are competitors' market shares? Market share can be measured in many ways. Usually for hospital systems, it is measured by the percentage of patients with a certain disease who walk through their doors versus a competitor's. For example, if there are 1000 open-heart cases done per year in city A and one of the hospitals handles 400 cases, then they have 40% market share of the open-heart market. The benefit of knowing competitors' market share is that if a hospital is thinking of adding a new clinical service product line, and one of their regional competitors already has 70% of that market, it is likely they will have more difficulty and expense reaching their target market share. Entering a new market with many competitors with small market shares is easier than entering a new market where there are two or three dominant competitors. For example, if a hospital is looking to purchase the new tele-medicine eICU technology to monitor and manage ICU patients from rural hospitals, then knowledge of other those hospitals' current capabilities and services is very important. If a hospital can capture the ICU management business of the surrounding rural and community hospitals before its competitors, then they will have captured ICU market share and the potential revenue that goes with it. A close look at competitors' market share information in the 3Q report is essential to help make an informed technology decision. Unfortunately, obtaining market share information for the 3Q analysis can be problematic because this hospital information is usually confidential. Information may be available from government websites such as Center for Medicare and Medicaid Services (CMS) (`www.cms.hhs.gov`). Look for service line statistics given in the number of beds, number of bypass surgical cases, etc.

Can we increase our market share? One way hospitals can define their business strategy is by looking at market shares in their various service lines. If a new technology will increase a service line market share, then this would be a compelling reason to acquire the new technology. The opposite also holds true.

CHAPTER 4

Question #3: Is It Worth It?

Figure 4.1: 3Q Method Map – Question #3: Is It Worth It?

This last question is meant to identify the financial gain or loss of adopting a specific new technology. Keep in mind that this analysis may be done for a number of new technologies simultaneously so their financials can be compared and ranked. This comparison more clearly illustrates which technologies are financially prudent to acquire. To a greater degree than the other two key questions in the 3Q Method, answers to this last question require a statistical analysis, best performed using spreadsheet software such as Excel.

It will be helpful for the reader to refer to the Appendix and read the information contained in "How do hospital and clinicians get paid?" to gain a basic understanding of hospitals and physician reimbursements in the United States. This will shed light on how the reimbursement system can either encourage or discourage the purchase of new technology.

The standard, accepted way of demonstrating the financial viability of adopting or purchasing a new technology is to develop an incremental financial projection of the yearly revenues and expenses associated with it. The Net Present Value (NPV) is the financial return on an investment (ROI) over a specific period of time, taking into account revenues and expenses associated with the new technology and the cost of money. Some institutions calculate NPV over shorter or longer periods, but for 3Q purposes, a five-year NPV projection will be used[1].

It is important to emphasize that this five-year projection is an "incremental" analysis. This means that only the revenue and expenses directly associated with the new technology are taken into account and not the entire hospital's or company's revenues and expenses. This greatly simplifies the task to answer the question, "Is It Worth It?"

A five-year projection includes the following:

- *Incremental revenues* each year due *just* to this technology.

---

[1] NPV is an Excel function: *NPV(rate,value1,value2, …)*. Note: "rate" is usually approximated by the inflation rate, with 5% used as an example.

- *Incremental costs (expenses)* each year for *just* this technology.

- Calculated cash flow each year (revenues minus costs).

- Calculated 5-year accumulated cash flow in one number(NPV)[2].

There are two suggested formats for financial projections – one for company cases and one for hospital cases. The concept is similar for both, but the specific components of revenue and expense differ.

4.1   COMPANY SETTING

Below is a sample spreadsheet that might be used for a company analysis. There are five columns for the annual figures. Here is an explanation of the categories on the left side of the spreadsheet:

The revenue from this new technology is a function of the average selling price (ASP) minus the product unit cost (manufacturing cost). *"Cost"* is used to represent what it costs the company to manufacturer the new product. *"Price"* means the money the customer pays to purchase the product. Thus, the unit cost and unit-selling price must be established prior to going to market. These price estimates usually can be obtained from companies dealing in similar products. If desired, these two amounts can be varied over the five-year period as in the sample table. It is not unusual for the product unit cost to decrease over time due to cost reduction efforts. Average unit selling price may decrease over time due to increased competition.

Income represents revenues from the new product. The number of units sold per year must be estimated. Total yearly sales ($) are then calculated by multiplying the number of units times the unit average selling price. The Gross Margin figure is calculated by first subtracting the Product Cost from the unit ASP, then multiplying by the number of units sold. "Margin" used here represents the net difference. Different accounting systems use different words for this, e.g., profit.

If a company must expend some engineering resource to integrate the new technology into their product line, which could take one or more years, sales revenue during this period could be zero. This should be reflected in the five-year projection as illustrated in the example below. If there are associated disposable products sold with each unit, then a separate spreadsheet must be created and merged with the new product spreadsheet to determine the total income.

Operating Expenses for companies in the medical device or systems businesses typically fall into two categories: engineering and marketing costs specific to the new product. Manufacturing costs are already included in the product unit cost numbers. Engineering figures require some estimation of the following costs: the number of engineering man-years needed to bring

---

[2]NPV is an Excel function: *NPV(rate, value1, value2, …)* Note: "rate" is usually approximated by the inflation rate, with 5% used as an example.

the product to manufacturing, multiplied by the cost per year of an engineer. For the base case, $150,000 per year is a reasonable engineer's salary, including benefits. Marketing costs should include estimates for advertising campaigns and publications. Research into similar industry figures will help to more accurately predict the engineering and marketing costs.

Operating Profit is the Gross Margin amount minus the Operating Expenses. The spreadsheet will automatically calculate the NPV over the five years.

The potential revenue impact of disposables associated with the new technology was mentioned in the section, "Can We Win?" The existence of disposables necessitates a 5-year projection of the disposable sales. This adds an additional variable on this spreadsheet projection – accumulated unit sales. Knowing the average rate of disposable usage per unit and the cost per disposable will allow the calculation of yearly disposable revenue. To calculate the accumulated unit sales for any given year, all previous years' incremental unit sales must be totaled and to that only half of the present year's unit sales should be added. This is necessary to account for the fact that not all units will be installed for the entire year. Using only half of the present year's units will reasonably approximate the disposable usage for this year's units. The calculation of accumulated units in any given year x is the following:

Accumulated units year $x = \text{units}_x/2 + \text{units}_{x-1} + \text{units}_{x-2} + \text{units}_{x-3} + \ldots\ldots + \text{units}_{x-n}$

Below is an example 5-year projection of a company that sells boxes that include a disposable kit used for each patient. Note the significant increase in operating profit because of the accumulated installations of boxes.

4.2   HOSPITAL SETTING

The incremental financial analysis for a hospital setting is a little more complicated. Although the question, "what is the financial benefit of acquiring this technology to the hospital?" is straightforward, healthcare accounting is extremely complex. Don't be discouraged by this known fact. Here are some helpful tips.

Hospital incremental income can come from two main sources: adding new services and reducing existing expenses. Because adding a new hospital service is similar to adding a new product to a company, the focus here is how to examine hospital savings projects.

Savings can be classified in terms of "hard" dollars and "soft" dollars. "Hard" dollars are actually cash you can take to the bank. For example, if a hospital decided to buy new technology ECG electrodes at a 15% price reduction, then this saving would be considered "hard" dollars. The hospital's financial statements could reflect these savings in real dollars – it's money that exists somewhere and can be put in the bank or used to pay for other goods. "Soft" dollars are usually represented in the form of more efficient workflow savings. What some new technology will do is to allow the staff to reduce their care procedures and at the same time, hopefully, raise the quality of care another notch. For example, if a new technology saves an ICU nurse five minutes for a certain

Table 4.1: Example of a 5-year NPV for a company technology adoption.

	Year 1	Year 2	Year 3	Year 4	Year 5
Product Unit Cost	$2,700	$2,700	$2,200	$2,200	$2,200
Unit Ave Selling Price(ASP)	$11,000	$11,000	$11,000	$9,800	$9,800
Income					
Sales (units)	0	270	670	1700	2200
Sales ($) @ASP	$0	$2,970,000	$7,370,000	$16,660,000	$21,560,000
Gross Margin	$0	$2,241,000	$5,896,000	$12,920,000	$16,720,000
Operating Expenses					
Engineering Costs	$450,000	$150,000	$30,000	$30,000	$30,000
Marketing Costs	$125,000	$70,000	$10,000	$10,000	$10,000
Operating profit	$(575,000)	$2,021,000	$5,856,000	$12,880,000	$16,680,000
				NPV=	$30,009,745

Table 4.2: Example of a 5-year revenue projection of product with a disposable.

Five Year Disposable Kit Projection -- Company
Box with associated disposable kit

	Year 1	Year 2	Year 3	Year 4	Year 5
Disposable Kit Cost	$5	$5	$5	$5	$5
Disposable Kit ASP	$35	$35	$35	$35	$35
Income					
Sales (boxes)	0	270	670	1700	2200
Accumulated sales (boxes)	0	135	605	1790	3740
Kits used per box per year	0	300	500	600	700
Disposable Kit Sales (unit)	0	40,500	302,500	1,074,000	2,618,000
Kit sales ($)	0	$1,417,500	$10,587,500	$37,590,000	$91,630,000
Kit Gross Margin	0	$1,215,000	$9,075,000	$32,220,000	$78,540,000
Operating Expenses					
Engineering Costs	$100,000	$30,000	$30,000	$30,000	$30,000
Marketing Costs	$-	$-	$-	$-	$-
Operating profit	$(100,000)	$1,185,000	$9,045,000	$32,190,000	$78,510,000
				NPV=	$96,790,435

bedside procedure, the soft dollars saved are the nurse's salary for those five minutes. These are "soft" dollars because they don't appear on the hospital's financial statements. However, soft dollar savings are still worthy of including in the 3Q report. The challenge is to see if soft dollars can be turned into hard dollars because *only hard dollars should be used on the five-year projection.*

Of special importance on this topic is the patient length of stay (LOS), because this is one of most commonly recognized financial outcome parameters in healthcare. One of the first questions in determining potential incremental income or savings is, "Does the new technology reduce LOS while maintaining or improving the quality of care?" This is one of the key questions asked by hospital administrators and enterprise management. Many times a new technology will lead to better patient treatments that will lead to shorter LOS, so it is the 3Q assessor's challenge to find these consequential chains-of-events that result in hard dollar savings. To include a LOS improvement in a 3Q five-year financial projection requires a peer-reviewed article(s) stating LOS improvement, or credible expert opinions.

A reduction in LOS will allow the hospital to accept more patients per year, increasing patient flow through the emergency room, ICU's, medical/surgical floors and clinics. Each incremental additional patient can contribute to the yearly hospital revenue. The Center for Medicaid and Medicare Services (CMS) pays the hospital a lump sum for each Medicare patient that is admitted. After expenses, this additional patient net income is something the hospital can take to the bank. Not all patients are not covered by CMS – only those over age 65, and those with minimal incomes. For more information and statistics on this, refer to the Appendix – "How do hospital and clinicians get paid?"

Here are some examples of possible incremental income or savings areas for the three different types of technologies that the 3Q Method covers: a) measurement devices, b) therapeutic devices and c) clinical information systems.

Hospital incremental expenses are more straightforward. Some of the more common expense categories accompanying new technology purchases are listed below. All of these expenses should be researched and documented in the 3Q final report.

Capital expense: This is the purchase price of the new technology. It can be paid for all at once, over time, or leased. In any event, this expense must be represented on the five-year projection.

Disposable expense: Many new technologies will need some disposable items, and this expense must be represented on the projection.

Training expense: Some level of staff training is typically necessary with new technology. Assume that the staff will need to complete their training outside of their regular clinical hours, so they will need to be paid for this extra time. Estimate the number of staff to be trained, hours of training time per person, and their hourly rate to arrive at a total expense. Any trainers and their material not covered with the initial installation should be included as a part of the training expense also.

New Technology	Possible Benefits
a) Measurement Devices	LOS reduction resulting in more patients/yr Less disposable items used Reduction of lab tests needed Reduction in imaging tests Faster ER triage resulting in higher patient throughput Reduction of hospital admissions Reduction in ER visits
b) Therapeutic Devices	LOS reduction resulting in more patients/yr Law suit reduction* Less ventilator days Shorter outpatient therapy sessions resulting in higher patient throughput Lower infection rates
c) Clinical Information Systems	LOS reduction resulting in more patients/yr Lawsuit reduction Reduction in imaging tests Faster review of medical record leading to faster decisions, leading to shorter LOS Reduction in medication errors means shorter LOS and less lawsuits Fast availability of patient medical history shortens therapy decisions and reduces LOS, Reduction of unbilled services improves revenue Reduction in redundant lab tests

Table 4.3: Example of hospital savings/benefits from new technology.

Additional staff expense: New technology may require additional staff. Their anticipated annual salary and benefits should be added to the projection.

Lost revenue: Sometimes the installation of new technology will cause a current hospital service to be shut down or minimized during the transition. This is lost revenue for the hospital and must be calculated and recorded on the projection.

Correct presentation of NPV results in the 3Q report is essential. The following is a suggested slide format:

1. Slide #1 - Assumptions.

   (a) Revenue assumptions.

   (b) Cost assumptions.

2. Slide #2 – Five year projection.

   (a) Five-year projection of revenues and expenses.

   (b) NPV for the 5-year projection.

Companies and hospitals will usually compare NPV's from different new technologies to see which provides a better NPV. A positive NPV indicates that there is a positive ROI over the five years, and a negative NPV indicates there is a negative ROI, or loss, over the five-year period.

4.3 CONCLUSION

In an academic setting, the final 3Q report is the culmination of researching the answers to the three key questions. However, in real life, management teams review this information from the 3Q report as a first step to arrive at an assessment of the feasibility of the new technology. Typically, the 3Q report is used as the first look at new acquisitions. The 3Q analysis can be accomplished in a relatively short time and at a minimum cost, unlike some more sophisticated feasibility studies. The management options after reviewing the 3Q report are to reject the technology from further consideration or to accept the technology for a more detailed analysis. Although extremely valuable, the 3Q Method is by no means meant to be the single definitive technology assessment. The emphasis is always with the company or hospital's strategic plan as the guide for which direction to move with new technology.

The strategic plan, coupled with sound analysis, impacts the weighting of the responses to each of the 3Q items. For instance, if a new technology fails the "Is it real" test by containing soft science or no clinical trial data, it will usually be rejected without the information from the other two questions. If the data leads to a "No" response to the question, "Can We Win?" the technology is usually rejected on that basis alone. And as for the final question – it still may be "worth it" if a new technology with a negative NPV has enough pull through sales in the five-year projections to turn around the NPV.

Table 4.4: Five-year hospital incremental projection example—Purchase of a Hi-tech ventilator.

Five Year Incremental Projection -- Hospital
Hi Tech Ventilator Purchase

Assumptions:
1) 10 bed Pulmonary ICU
2) Ave patient LOS = 7 days
3) Hi Tech Vent. Saves 1 day
4) 14 additional patients/yr
5) Profit/patient = $6000
6) Ventilator price = $10,000
7) 30 RN's ($50/hr) & 10 Resp Tech ($50/hr) trained
8) Training takes 2 hrs

	Year 1	Year 2	Year 3	Year 4	Year 5	TOTALS
Incremental Income						
Incremental Patient Throughput	$84,000	$84,000	$84,000	$84,000	$84,000	$420,000
Incremental Expenses						
Capital Expense	$100,000	$0	$0	$0	$0	**$100,000**
Training Expense	$4,000	$0	$0	$0	$0	**$4,000**
Lost Revenue (if any)	$0	$0	$0	$0	$0	**$0**
Addition Staff Expense	$0	$0	$0	$0	$0	**$0**
Incremental Income - Expenses	-$20,000	$84,000	$84,000	$84,000	$84,000	**$316,000**

NPV= $264,628

Finally, the ultimate success of the 3 Question (3Q) medical technology assessment method rests with the individual's skill to ask the right questions, research the answers and distill complex information into a readily understandable, actionable communication to management.

For further information on creating the actual report, go to the to Appendix – "Elements of a Good Technology Assessment PowerPoint Report." This document provides some basic guide points for designing a professional 3Q report.

CHAPTER 5

3Q Case Study Example – Pershing Medical Company

5.1 PERSHING MEDICAL COMPANY (PMC) BACKGROUND

Pershing Medical Company is a medium-sized firm that presently sells a line of pulse oximeters with disposable finger and ear sensors worldwide. PMC was founded in 1987 by James Trufniew, Ph.D and Emma Nonnel, MD. The company employs 270 people – all of whom work out of their headquarters in Lodi, Wisconsin.

PMC has been doing well financially since 1992 when they turned a profit for the first time. Their recent financials are listed below in Table 5.1.

Table 5.1: PMC sales, costs and margin for 2004-2008.					
Year	2004 ($M)	2005 ($M)	2006 ($M)	2007 ($M)	2008 ($M)
Sales	95	105	125	145	155
Costs	75	80	105	115	120
Margin	20	25	20	30	35

PMC competes worldwide with three other firms in the pulse oximetry market. The competitors and their respective market shares are listed below in Table 5.2. PMC has been struggling to increase their market share over the past three years, but it has had limited success as can be seen

Table 5.2: Recent market shares in pulse oximetry market.			
	World Wide Market Share		
Company	2006 (%)	2007 (%)	2008 (%)
Johnson Medical	42	40	38
Dubai Medical	36	36	39
Pershing Medical (PMC)	**17**	**18**	**18**
Solar Medical	5	6	5

5.2   CURRENT SITUATION

PMC has just completed their five-year strategic plan. One of the plan initiatives is to enter new product lines in order to boast sales while continuing to push their pulse oximetry product line. They have been approached by a small start-up company, Nobis LLC, which claims to have invented a non-invasive cardiac output (CO) device. Nobis is interested in licensing their technology to PMC on a non-exclusive basis. Nobis has done both a clinical trial comparing their device with a gold standard device and a prospective, randomized clinical trial to demonstrate improved patient outcomes. These trials produced clinical data that were sent to the FDA as part of their 510(k) submission. The FDA has acknowledged receipt of their submission and is presently evaluating it.

Adam James works in Research and Development at PMC. His manager has asked him to do a preliminary evaluation of the Nobis technology. His report will be the first piece of information used to decide if Nobis' technology should be investigated in more detail for possible licensing or acquisition consistent with PMC's five-year strategic plan. Adam begins his analysis with the 3Q medical technology assessment method. The questions in bold below are those that Adam asks as the relevant framework for his 3Q analysis.

5.3   IS IT REAL?

What is the clinical problem solved by a non-invasive cardiac output (CO) device? Adam reviews the medical literature on the topic via the Internet and determines that CO is important for managing critical care patients throughout the hospital system: ER, OR, ICU and Medical/Surgical floors. Adam's approach is to visit several hospitals for first-hand observation and consult via phone a cardiologist, an intensivist, and a nurse.

From these informational sources, Adam ascertains that CO is one of the most important physiological variables for managing cardiovascular patients. He learns that to best manage a cardiovascular patient, the clinician would like to know CO, blood pressure (BP) and heart rate (HR). BP and HR are presently available in CHF clinics. The problem, the clinicians tell him, is that all of the present and reliable methods for measuring CO are invasive. The present method of invasively measuring CO is with a catheter fed into a vein, through the right atrium and right ventricle and into the pulmonary artery (PA). This procedure is labor intensive and very risky for the patient because of the potential for clots, punctures and infection. Adam's expert sources all agree that having a reliable, non-invasive CO (NICO) technology would be extremely beneficial. Adam carefully documents this information that verifies the clinical problem and its sources.

What is the projected clinical impact? Adam's clinical consultants tell him that cardiovascular disease is one of the top medical costs today, and they believe NICO will have a great impact on care, producing a new worldwide market for PMC. Adam realizes that NICO will not immediately replace all the invasive CO systems in use because Nobis' NICO does not provide all of the information the PA catheter does. Some of the very sick patients will still need a PA catheter. Of course, this all assumes that the Novis NICO device is as accurate as the PA catheter method.

Are there safety improvements? Adam realizes that the use of NICO instead of the invasive PA catheter may avoid catheter infections and other risks connected with its use. He researches several clinical articles that document the rate of complications of the PA catheter. They are significant.

What are the workflow improvements? Adam discusses this with the critical care ICU nurses who manage the present PA catheters and perform the CO measurements multiple times each day for each patient. They tell him that the catheter and thermal dilution bedside apparatus requires a high degree of skill to administer and monitor. There are plenty of opportunities for error in making a CO measurement. A simpler, non-invasive CO method would likely be less labor intensive and more accurate.

The Nobis NICO technology requires two pulse sensors to be placed on the patient, which is one additional sensor than the present convention. Adam believes that this will pose some additional set-up for the nurses. However, the value of continuous NICO monitoring, he believes, will convince the nurses of this beneficial trade-off. Other than the extra pulse oximetry sensor on the patient, the new NICO technology integrated into the existing bedside monitors will be almost transparent to the user.

Does the new technology create new problems?

The Nobis NICO requires two pulse oximetry sensors on the patient – one on the finger and one on the ear. Putting two sensors versus one on the patient will be added workload on the nurses. Adam notes this fact in his report since extra work for clinicians will diminish the attractiveness of the product.

Is there clinical data available? Adam sees that Nobis has done two clinical studies, and he must evaluate each separately. The first study is a gold standard comparison of the Nobis device with CO derived with the standard PA catheter using the thermal dilution method in an ICU setting. There were 22 patients and 42 cardiac output measurement points. Nobis funded this study, and it has *not* been submitted to a medical journal. Adam notes that there may be possible bias or conflict of interest in the self-funded study. Here is the corresponding Bland Altman plot, with FWCO as the Nobis device. PACO is the CO measured by the PA catheter.

Examining Figure 5.1, Adam observes that the bias = 0.02 l/min and the limits of agreement (bias ±2 standard deviations) = −2.33 l/min and +2.36 l/min. Looking at the bias and knowing from his Internet research that the normal adult CO is approximately 5.0 l/min, Adam calculates that a bias of 0.02 l/min is 0.02/5.0 or a bias of 0.4%. Checking with his clinician colleagues, he concludes that 0.4% bias is well within the clinical acceptance range. He now must see if the precision or limits of agreement are acceptable. He contacts the intensivist to see if a device that measures CO, with 95% of the readings between −2.33 l/min and +2.36 l/min of the actual CO value would be acceptable. The response from the intensivist is that this accuracy would be acceptable if the patient's CO is 8 l/min or higher, but since the normal adult CO is closer to 5.0 l/min and can be as low as 2.0 l/min for sick patients, −2.33, +2.36 l/min is *not* acceptable. An error of this magnitude could produce a wrong clinical decision. For example, in a sick patient with a CO = 2.5 l/min, a change of 0.75 l/min will cause the doctor to make a cardiovascular med change. A 95% confidence interval

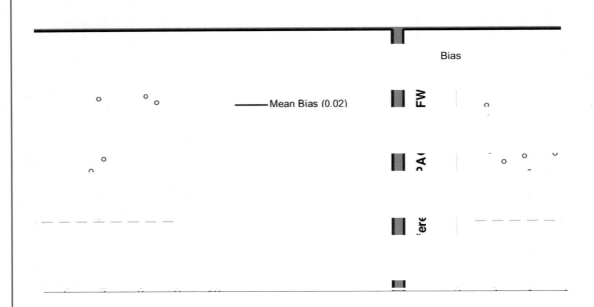

Figure 5.1: Nobis, Inc. Bland Altman Plot of Gold Standard Study.

of −2.33, +2.36 l/min will not give the clinician the accuracy needed to make these decisions, (i.e., a 0.75 l/min observed difference could be due to the unreliability of the measurement). Adam's cardiologist colleague confirms this opinion that the CO limits of agreements (2 SD) should be ±0.75 l/min at CO's below 5 l/min. Nobis does not meet this requirement.

Adam then analyzes whether Nobis has sufficient CO data points on the Bland Altman plot over the clinical range of CO. From discussions with his clinical colleagues and literature, he determines the clinical CO measurement range is 2.0 – 12 l/min. Inspecting the Nobis Bland Altman plot of Figure 5.1, he observes very few data points below 3.5 l/min and no data points over 8.5 l/min. Adam then concludes that the Nobis gold standard study fails in two areas: a) limits of agreement are too wide for clinical use, and b) the study did not have sufficient data points at the low and high end of the clinical CO range.

The Bland Altman results are not completely conclusive, but they certainly cast doubt on how well the NICO agrees with the gold standard. One advantage of the NICO, however, is the ability to use the technology in situations where the gold standard technology would be difficult or impossible to implement. While the NICO may have less-than-perfect accuracy relative to the gold standard, its continuous monitoring ability may still lead to better outcomes. To explore this possibility, Adam pushes on to look at the second study conducted by Nobis.

The second study was a two-arm, prospective, randomized clinical trial in an ambulatory congestive heart failure (CHF) clinic. The study hypothesis stated that adding NICO to the existing set of measurements would speed up the treatment and shorten the LOS of the patients. A local CHF clinic was chosen to do the study. In this unit, CHF patients present in acute failure and the purpose of the treatment is to return them to cardiovascular stability as quickly as possible and then discharge them back home. At the present time, CO cannot be accurately measured non-invasively and these patients typically do not have a PA catheter in place because of the risk. Consequently, the clinicians manage the patients without CO.

Only patients in acute failure were candidates for this study. All acute CHF patients who came to the clinic were automatically placed in the study, unless they opted out. No other exclusion criteria were used. The average length of stay for acute CHF in this facility has been historically 12.1 ± 4.5 hrs.

One arm of the study used the current assessment methods without CO and the other arm used the new NICO device to help guide therapy. Measured outcomes were LOS in the CHF Unit. The study was carried out until each arm contained 30 patients. Patient mean age and age range were equivalent in the two arms of the study. The results of the study are below.

Table 5.3: LOS Study results for the CHF use of NICO.			
	No NICO (n=30)	**NICO (n=30)**	**p value**
LOS	13.2 ± 4.1 hrs	11.3 ± 2.0 hrs	0.03
CO mean $\pm$ SD	4.6 ± 0.70 l/min	4.2 ± 0.53 l/min	0.16
CO range	$3.5 - 6.4$ l/min	$3.4 - 7.1$ l/min	0.13

The study indicated that there is a 1.9-hour reduction (i.e., $13.2 - 11.3$) in LOS with the NICO technology for assessing CHF patients in this unit. This difference is statistically significant since $p < 0.05$. Adam now asks the question, "Is this clinically significant?" To answer this, Adam researches the costs to treat a patient per hour. The clinic's average expenses are $70/hour for nursing care, $66/hour for meds and supplies, and $250/hour overhead for a total of $386 per hour per patient. Over the last 12 months the clinic had 566 patient visits. A 1.9-hour LOS savings per patient could mean a yearly savings of $415,104 (566 visits x $386 cost/hr x 1.9 hr/patient). Taking into account that this study was large enough to detect a clinically meaningful difference, these are positive results for this technology. However, it is only one study at one institution.

Are the clinical limits tested? After researching NICO's clinical application areas, Adam comes up with the requirements regarding clinical limits, detailed in the table below.

Are these study results applicable to the average CHF unit? Table 5.4 helps to answer the question that the data in the study unit is representative of the typical CHF unit. Adam now feels comfortable including the study results in his 3Q report. He was not successful in finding any other

Table 5.4: Clinical limits requirements *vs.* NICO study parameters.

Clinical item	Clinical requirement	NICO clinical study	Comment
Age range	18 – 90 yrs for general population 60 – 90 yrs for CHF Clinic population	62 – 87 yrs	Close enough
Weight range	100 – 275 lbs	105 – 236 lbs	Close enough
CO range	2.0 – 12 l/min	3.0 –8.0 l/min	Not representative of typical CHF Unit population
Typical med therapies	Inotropic, chronotropic & vasoactive meds, diuretics	Most study patients treated with Inotropic, chronotropic & vasoactive meds, diuretics	Similar to other Units
Typical other therapies, e.g. ventilator, pacemakers	Not applicable in CHF Units	Not applicable	Not applicable
Common co-morbidities	Ventricular arrhythmias; valvular stenosis & regurgitation; pulmonary hypertension	Multiple cases of each co-morbidity within both study arms	Similar to other Units

clinical studies using this NICO technology, so he will have to rely on this initial study to indicate the clinical benefit of this technology.

What are the training requirements? After interviewing three nurses from three different hospitals, Adam believes that the additional training for this new NICO technology will be minimal. Instructing the nurses to place an additional pulse oximeter on the patient requires no formal training, simply a procedural change. Instructing them on how to clinically interpret continuous CO measurements will take about one hour. Adam concludes that the training requirements for the new technology have no negative impact on this 3Q assessment.

Is the technology based upon real scientific principles? Adam has spent some time researching the basis for the NICO algorithm. He believes this is the key element of the method, and from his past experience in cardiovascular measurements, the NICO algorithm could be the weak link. The following documentation is provided by the NICO inventor:

> "Measurements are calculated on beat-by-beat basis, with time averaged updates displayed. Three separate pieces of information form the building blocks of the Cardiac Output measurement: Pulse Transit Time, Volumetric Pulse Contour, and Patient Demographics. Pulse Transit Time and Volumetric Pulse Contour are calculated through ECG electrodes and pulse oximetry sensors(s). Two-lead ECG electrodes are placed on the torso. Pulse oximetry sensors are placed on a digit of each hand."

From the literature, Adam discovers that the pulse oximetry's pressure contour during systole is proportional to the amount of blood pumped out by the left ventricle, i.e., cardiac stroke volume (SV). The proportionality "constant" is the compliance of the arterial tree. Compliance is the change in arterial volume divided by change in arterial pressure (C = Δ Volume/Δ Pressure). So Adam concludes that one could determine SV by knowing the compliance of the arterial tree. Knowing SV, CO can be calculated by multiplying heart rate (HR) by SV where SV is Δ Pressure/C. These are well known clinical relationships.

$$CO = HR \times SV = HR \times (\Delta Pressure/C)$$

In performing this calculation, one would have to measure or estimate arterial compliance. Adam finds that measuring arterial compliance directly is extremely difficult on humans and, hence, is not presently done in a clinical setting. According to the inventor's information, Nobis estimates compliance for each heartbeat by measuring pulse transit time (PTT) from the ECG waveform and the pulse oximetry waveform and combining it with the patient's demographic information, e.g., height, weight and gender. The exact algorithm is not spelled out in the inventor's information because it is proprietary. However, Adam also learns that arterial compliance can be highly non-linear depending on a range of arterial pressures, depending on a several clinical factors such as age and vasoactive medications. This throws up a caution flag in Adam's mind. Can PTT and demographics account for all the compliance non-linearities? If the patient suddenly starts or stops vasoactive medications, how will the Nobis compliance algorithm adjust for this? Adam further determines that the Nobis clinical study does not adequately test these common clinical ranges of conditions. He also knows that the calculation of PTT is problematic under real clinical conditions. PTT is typically in the range of tenths of a second. Good, consistent PTT accuracy is needed for each beat if SV calculation is to be accurate. Measuring PTT accuracy to within tenths of a second for patients moving around in bed is difficult at best. Additionally, he learns that several other physiological variables not yet considered effect PPT. From this analysis, Adam is concerned about the lack of clear physics or physiological principles involved in the NICO algorithm. He determines that much more clinical testing is necessary for this algorithm to resolve these gray areas and be robust enough for clinical use.

Adam turns his attention on summarizing his 3Q analysis for his manager. Below is the PowerPoint slide that Adam created based upon his "Is it real" analysis.

Adam now turns his research to the next 3Q assessment question.

5.4   CAN WE WIN?

Is the new technology a strategic fit for us? Adam knows from PMC's strategic plan that the company is looking for new products and markets with greater than 10% growth rate. Although NICO would be a new product for PMC, part of the NICO technology that involves pulse oximetry is very familiar to PMC. PMC is also familiar with cardiopulmonary medicine because of their pulse oximetry products. Having a CO product is consistent with the company's present clinical knowledge base.

NICO serves the CHF clinic market, which is a new market for PMC. Their pulse oximetry products do serve general clinic markets, but PMC has never created a targeted marketing program for the CHF clinics. Adam believes that PMC could create a marketing program for this new market because NICO would be a PMC proprietary product, and NICO is desperately needed. Adam creates a market map of the NICO strategic move to help his manager understand his rationale. See Figure 5.2 below. Using this map of PMC's product portfolio, management can visualize where the Nobis technology is taking them and whether it is consistent with the PMC strategic plan. Other possible strategic moves also become evident using this four-block map as a discussion point.

Will it improve PMC's competitiveness? Since PMC would be the only vendor with a NICO product, it would definitely give PMC an advantage in the CHF clinic market.

Will it give us a leadership image? Since the ability to measure CO non-invasively is a breakthrough technology, Adam feels this will greatly enhance PMC's leadership image in the cardiopulmonary arena.

Will it leverage sales of PMC's present products? Adam believes that clinics that have never purchased PMC pulse oximetry previously will now buy their NICO product and thus "pull through" their pulse oximetry products. There are actually two pulse oximeters in each NICO product so, indeed, PMC pulse oximetry sales should increase because of NICO.

Will NICO cannibalize any PMC's existing products? Adam believes the NICO product will not obsolesce any existing PMC product. The present PMC oximeters will be viewed by cus-

Table 5.5: Sample summary slide for "Is it real?"

Nobis Co NICO Technology Analysis
12/10/09 A. James

Solve a real clinical problem?	**Based upon real scientific principles?**
CHF treatment large health expense – potentially large ambulatory & in-patient markets CO, BP & HR are key to CHF management CO cannot be presently measured non-invasively so clinicians have to estimate CO for ambulatory patients NICO would greatly enhance clinical assessment of CHF	NICO accuracy dependent on estimating arterial compliance from patient's demographics & PTT – little existing literature to support this -- has loose physiological or physics basis Inherent error in PTT measurement will overwhelm needed accuracy making algorithm possibly unreliable CO algorithm not tested over clinical entire CO range in Nobis study
Clinically tested?	**Conclusion**
Gold standard comparison study (n=22) ◦ Funded by Nobis ◦ Precision <u>unacceptable</u> (2.4 l/min vs. required 1.25 l/min) ◦ Limited CO range tested <u>Insufficient data points at low and high end of CO values</u> Two arm randomized study ◦ n = 60 ◦ LOS in CHF Unit reduced by 1.9 hrs (14%) (p = 0.03) – could save $415K/yr ◦ Limited CO range tested	Real cardiovascular opportunity! Clinical studies did not test full CO range (funded by Nobis) Precision too large -- clinically unacceptable Algorithm not based upon recognized science

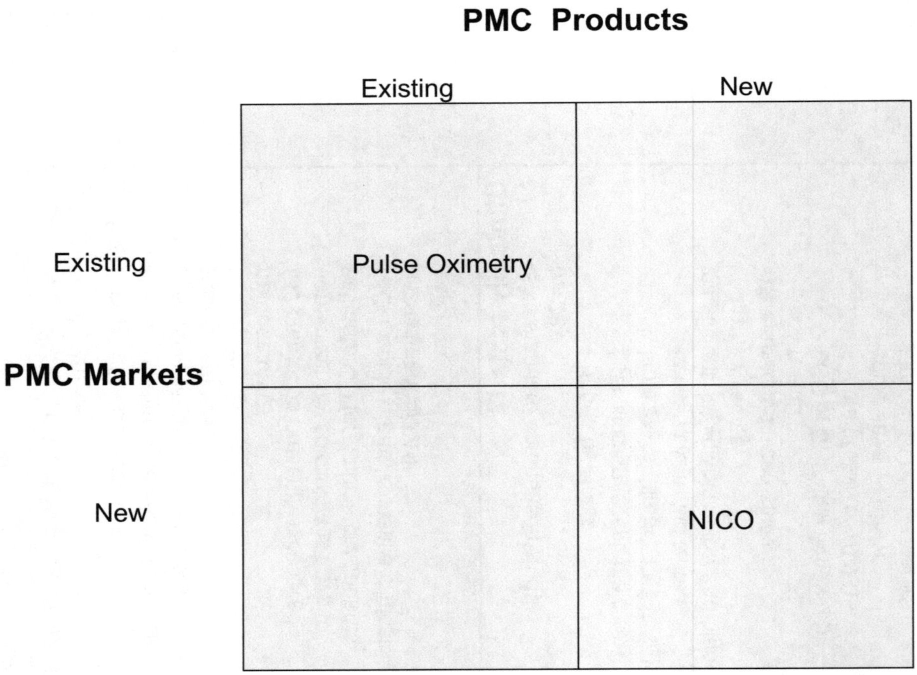

Figure 5.2: PMC Product-Market Map of the New NICO Product in Relation to Existing Pulse Oximetry Products.

tomers as a separate product and will continue to be relevant in their respective markets. In the clinic markets where both oximetry and CO are needed, the new NICO product be purchased in place of one of the existing oximetry products. This is a positive outcome since, effectively, the customer will be buying a higher price product from PMC.

Do we presently sell to target customers? Even though PMC does not presently target cardiologist in the CHF clinics, PMC knows the cardiopulmonary market in general, and its name is well known in these arenas. The PMC sales force could be additionally directed to the clinics to push the NICO product with some additional training. This marketing effort would be accelerated by a direct mail and ad campaign in cardiology journals. Adam feels comfortable that PMC can effectively target these customers. PMC uses distributors to sell their products in Europe and Asia, but because of the operation simplicity of NICO, the dealer training could be easily accomplished by PMC.

Does the product include disposables? Adam realizes that this is a very important question, because disposable products can greatly increase the revenue of a product. In fact, at this time PMC derives 55% of its sales from the disposable pulse oximetry sensors. The proposed NICO product has two pulse oximetry sensors along with three disposable ECG electrodes per patient. This represents a significant disposable revenue stream for PMC.

Can PMC service it? Adam knows the importance of effective customer service in the pulse oximetry business. He views the service requirements of NICO and pulse oximetry to be similar in terms of phone response time, repair turnaround, availability of parts, etc. He sees no problem with merging the NICO product into their current service system.

Can we use the present training staff? Effective user training is important if PMC expects that NICO will become a standard of care in the CHF clinics. Right now, PMC uses its sales force to install and train the customer in the United States; dealers perform the offshore training. Adam believes that the cardiologist customer will need no training on how to manage his/her patient with CO since that is a well-recognized clinical variable. Therefore, teaching the customer how and where to place the three ECG electrodes should be an easy addition to the sales' customer training toolkit.

What are regulatory requirements? Nobis' NICO product is a 510(k) FDA device, as are PMC's pulse oximetry products. Hence, NICO does not require a new FDA regulatory approach. Adam is concerned, however, that the Nobis clinical data submitted to the FDA are weak and, probably, will have to be augmented with additional clinical testing.

How large a marketing and education effort is it to move customer to this technology? Because NICO does not introduce a new clinical variable or method, Adam believes the rate of increase of NICO sales will not be limited by a technology learning curve.

Who are the competitors? Right now, PMC is the number three pulse oximetry vendor with an 18% market share. The company's strategic plan is to penetrate a new market in the next three years and maintain or increase the present products' market share. Currently, there is no NICO competition, so PMC will have 100% market share at the onset. The issue is how fast the competition will enter the NICO market. Nobis' patent coverage will be an important issue in restricting fast-following competitors. Adam will recommend that PMC's patent attorneys scrutinize these patents to see how effective the barrier is to competition.

Do we have the clinical understanding? Because PMC presently sells a cardiopulmonary product that measures arterial oxygen saturation and heart rate, Adam believes the step-up in company clinical expertise will be minimal.

Do we have the right R&D expertise? If PMC takes over continued development of the NICO product from Nobis, it will have to learn the basis for the derivation of CO from the two-pulse oximetry and ECG signals. PMC has a competent group of software engineers that do the development for the pulse oximetry algorithm. From what Adam has seen, the NICO algorithm is no more complicated than the current pulse oximetry algorithm and is within PMC's R&D capability.

Do we have the manufacturing capability? The addition of the ECG channels is the only new aspect of the NICO product compared to PMC's present oximetry products. ECG technology has been around a long time, and its manufacture is relatively straightforward.

After this 3Q "Can We Win?" analysis, Adam feels there is definitive synergy with PMC's products and markets for NICO.

5.5   IS IT WORTH IT?

Will ROI/NPV meet our standards? Adam knows that his management will require a five-year incremental revenue projection if the Nobis technology is added to the PMC product line. He starts first by examining what costs will be involved in integrating NICO into PMC.

1) Engineering costs – the existing NICO hardware and software design was really a prototype design and not final product design. So PMC will have to spend about one year doing the final design based upon the Nobis prototype. Adam figures this will take two hardware and two software engineers. He will use PMC's historical cost for all engineers of $100,000/year, which includes benefits. He calculates he will need all four engineers for the first year, two engineers the second year and one engineer each year after that.

2) Marketing costs – these costs will mainly fund journal ads and a direct mail campaign. From company historical data, he estimates that ads will cost $100,000 for the first sales year and $50,000 a year for the next two years. The direct mail campaign will consist of a mailing prior to the next two annual American Heart Association meetings and another mailing just before the next two annual American College of Cardiology meetings. He estimates that each mailing will cost $25,000.

3) Sales force training – Adam intends to add an extra four-hour session to the annual sales meeting to cover the new NICO product training. He estimates that the incremental cost for this will be the cost for a "NICO sales starter kit" that will be given to each of the 100 sales people. Each kit will cost PMC $150, for a total of $15,000.

4) Manufacturing costs – incremental start-up manufacturing costs will only be incurred in the first year of sales, and then it will be merged into the standard product cost accounting system. The start-up costs will include design of the new production area and generation of manufacturing documentation. Adam estimates that this will take one manufacturing engineer one year at $100,000/year. Tooling will be needed for the NICO cabinet and he estimates that will cost $25,000.

Besides the cost piece of the five-year projection, Adam must estimate some individual product costs, price and sales figures over the five years.

1) NICO product cost – at PMC the term "product cost" refers to the recurring manufacturing charges to build the product. It includes parts and labor. Adam estimates that the NICO product will contain three circuit boards @ $300, a cabinet @ $150, two pulse oximetry sensors @ $90 and three cables (ECG and two pulse oximetry cables) @ $50. This will bring the total product cost to $1380.

2) NICO product selling price – selling price refers to the listed price that will be quoted to the customer. The average selling price (ASP) is the price quoted after all discounts have

been given. For simplicity, Adam will assume the selling price is the ASP for the five-year incremental projection. PMC typically arrives at a minimum selling price by multiplying the product cost by three. This is a rough "mark-up" approximation that covers the recurring manufacturing product cost and PMC's minimal margin. So the minimal product price would be $4140 (3 x $1380). However, Adam believes that PMC can obtain a higher-than-minimum margin since NICO has no competitors at this time. He estimates that PMC could price the NICO at $5500.

3) Unit sales –the US market for NICO are the 4000 CHF clinics, other generic clinics, hospital emergency departments and hospital medical/surgical floors. For this 3Q analysis, Adam decides to include only the USA CHF clinics since he is more confident of his estimates in this arena. During his report presentation, he can mention the other NICO markets and offshore sales as upside opportunities. Adam estimates that each CHF clinic will eventually buy three NICO units, assuming NICO will be come a standard of care. Therefore, he believes that the US unit market will be 12,000 NICO units (4000 × 3). With an aggressive ad and direct mail campaign, and the assumption that there will be no competition for at least four years, PMC can potentially ramp up their unit sales quickly over the first four sales years.

Besides the equipment or "box" sales above, the NICO product would have a disposable "kit" that would be used for every patient. Adam does a separate projection for the disposable kit.

The "Is it worth it?" analysis illustrates that there would be a positive, five-year NPV of $23M for the equipment, and a NPV of $97M for the accompanying disposables, for a total NPV of $120M over five years. Adam is very excited about this NPV. It would also take PMC into a new market, which is one of their strategic goals. NICO sales also contribute a higher margin than PMC's existing pulse oximetry products.

Adam constructs a final summary slide of his 3Q analysis for his manager. (Note that Adam would present more detailed slides for "Is It Worth It?" and "Can We Win?" as he did for "Is It Real?" presented earlier. The following slide collects the summaries and conclusions from detailed presentations of all 3 questions. This is called a four-blocker slide because it uses that basic layout to present the information.

Adam now feels his analysis is complete, and he will first present his findings to his manager. After this discussion, he will prepare a slide presentation for presentation to the PMC new technology review committee.

Table 5.6: PMC five-year financial projection for the NICO product.

Five Year Projection -- PMC
Adam James

	Year 1	Year 2	Year 3	Year 4	Year 5
Product Unit Cost	$1,380	$1,380	$1,380	$1,380	$1,380
Unit Ave Selling Price(ASP)	$5,500	$5,500	$5,500	$5,500	$5,500
Income					
Sales (units)	$0	$500	$1,200	$2,200	$3,200
Sales ($) @ASP	$0	$2,750,000	$6,600,000	$12,100,000	$17,600,000
Gross margin	$0	$2,060,000	$4,944,000	$9,064,000	$13,184,000
Operating Expenses					
Engineering Costs	$400,000	$200,000	$100,000	$0	$0
Marketing Costs	$0	$150,000	$100,000	$50,000	$0
Manufacturing Costs	$125,000	$0	$0	$0	$0
Operating Profit	-$400,000	$1,710,000	$4,744,000	$9,014,000	$13,184,000
				NPV=	$23,013,963

Table 5.7: PMC five-year financial projection for the NICO disposable kit.

	Year 1	Year 2	Year 3	Year 4	Year 5
Disposable Kit Cost	$5	$5	$5	$5	$5
Disposable Kit ASP	$35	$35	$35	$35	$35
Income					
Sales (boxes)	0	500	1200	2200	3200
Accumulated sales (boxes)	0	250	1100	2800	5500
Kits used per box per year	0	300	500	600	700
Disposable Kit Sales (unit)	0	75,000	550,000	1,680,000	3,850,000
Kit sales ($)	0	$2,625,000	$19,250,000	$58,800,000	$134,750,000
Kit Gross Margin	0	$2,250,000	$16,500,000	$50,400,000	$115,500,000
Operating Expenses					
Engineering Costs	$100,000	$30,000	$30,000	$30,000	$30,000
Marketing Costs	$-	$-	$-	$-	$-
Operating profit	$(100,000)	$1,185,000	$9,045,000	$32,190,000	$78,510,000
				NPV=	$96,790,435

Table 5.8: Summary slide of the 3Q PMC report.

Nobis Company's NICO Technology Analysis
12/10/09 A. James

Is it real?

Real cardiovascular opportunity!

Clinical studies did not test full CO range (funded by Nobis)

Precision too large -- clinically unacceptable

Algorithm not based upon recognized science

Can we win?

NICO breakthrough technology – no present competitors -- enhance leadership image

NICO emerging market with >25% growth rate -- new PMC market

Nobis 510(k) submission weak – will require more clinical data

Worldwide distribution through present direct sales and dealer network

Is it worth it?

Favorable equipment NPV = $23M

Favorable disposable NPV = $97M

Recommendation

New, high growth market for PMC with unique NICO product

Favorable NPV of $120M (box & disposables)

Only one clinical study - unacceptable clinical test data

High potential product with weak clinicals – consider financing further clinical studies

APPENDIX A

Health Care Technology Assessment Sample Class Syllabus

Class Date	Class Agenda
	Table A.1: Example of a Typical Class Syllabus.
1/20	Intro to the 3 Question (3Q) Assessment Method IS IT REAL? -Types of clinical studies (MAARIE)
1/27	IS IT REAL? - Gold Standard comparison -- Bias & Precision; Bland-Altman analysis ***Assign #1 Report – Is It Real?***
2/3	IS IT REAL? - Bland Altman - Intro to Sensitivity & Sensitivity
2/10	***Team Report #1***
2/17	IS IT REAL? Testing a Test - Sensitivity & S
2/24	IS IT WORTH IT? - 5 yr ROI ***Assign #2 Report – Is It Real? & Is It Worth It?***
3/3	IS IT WORTH IT? - 5 yr ROI
3/10	***Team Report #2***
3/17	**No Class – Spring Break**
3/24	*Field trip to ICU – ICU technology assessment discussion*
3/31	CAN WE WIN? ***Assign Report #3 – All 3Q***
4/7	CAN WE WIN?
4/14	***Team Report #3***
4/21	ALL 3Q AREAS - Case studies ***Assign Report #4 – All 3Q***
4/28	ALL 3Q AREAS Venture Capitalist's Perspective
5/5	***Report #4 Presentations***

APPENDIX B

How Do Hospitals and Clinicians Get Paid?

It is important to understand how hospitals and clinicians are reimbursed as it relates to the 3Q Method of assessment. In some cases, new technology will increase reimbursement to hospitals, and in other cases, it will have no effect or a negative effect. Thus, the financial motivation of hospitals to buy new technology is greatly influenced by the different reimbursement methods. This payment system involves three key entities: private insurance companies, CMS (Center for Medicare and Medicaid) companies, and providers. Several facts are key to project whether a new technology will be financially beneficial to a hospital. This is important when a hospital asks the question, "Is It Worth It?"

Below is an explanation of the three general methods for reimbursing hospitals for *hospitalized* patients. Since this system is complex, only the high points will be described here.

1) *Reimbursement from CMS* (patient is 65 years or older) – CMS uses the DRG (Diagnostic Related Group) system to reimburse hospitals for a patient's stay. Each diagnosed disease has been assigned a DRG number that has a defined dollar reimbursement. Once the diagnosis has been made and a DRG number is automatically assigned the hospital is reimbursed that set amount for that patient's stay. The hospital then uses that money to care for the patient during the hospital stay. If the patient can be treated and discharged quickly, the hospital may make money. If they allow the patient to stay excessively long, they will loose money. This system motivates the hospital to minimize the length of stay (LOS).

2) *Reimbursement from private insurance companies* (patient is less than 65 years) – Usually, insurance companies are contracted by companies to cover their employees. The insurance companies typically cover all the hospital expenses while the patient is in the hospital. Hence, the hospital has little motivation to discharge the patient quickly other than the pressure from the insurance company. Notice this is a very different motivating factor compared to (1).

3) *Reimbursement from the patient* (patient is less than 65 years) – If a patient does not have health insurance, the patient is billed directly for all hospital expenses, i.e., fee for service.

The following two diagrams illustrate the first two cases. The third case is straightforward and need not be illustrated.

Table B.1: Reimbursement from CMS.

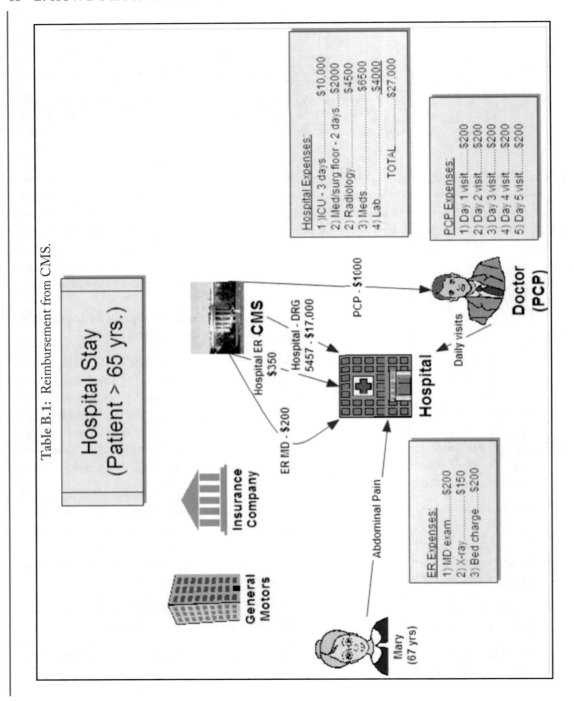

This is the reimbursement flow for a patient (Mary) who is over 65 years. The boxes include expenses. Note that emergency room (ER) reimbursement is not covered by DRG's since the patient has not been admitted to the hospital. ER costs are reimbursed via fee for service. PCP = primary care physician. Note that General Motors and the insurance company are not involved in this case.

Table B.2: Reimbursement from private insurance companies.

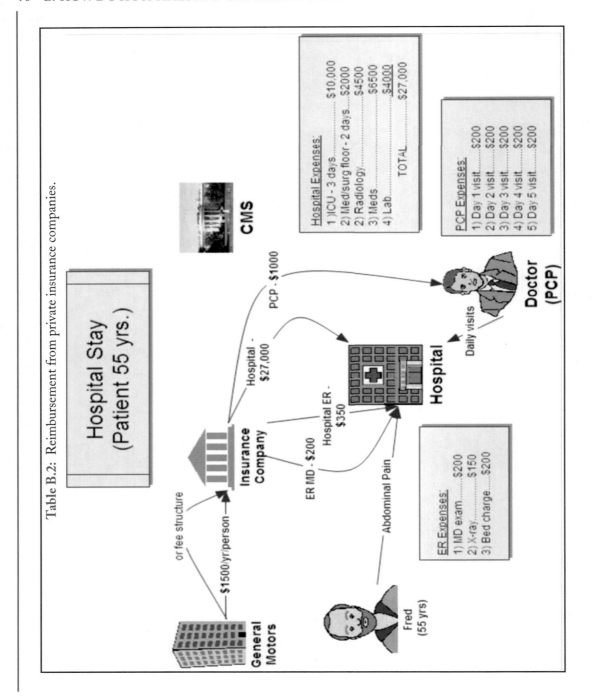

Note that emergency room (ER) reimbursement is not covered by DRG's since the patient has not been admitted to the hospital. ER costs are reimbursed via fee for service. PCP = primary care physician. In this case, her employer (General Motors) has established her coverage and/or fee structure with the insurance company.

Table B.3 summarizes the main hospital outcome variables that are possibly affected by the purchase of a new technology for both patients under and over 65 years. It also rates the appeal of this technology to the hospital administration for both the private insurance and DRG/Medicare cases discussed above.

Table B.3: Appeal to a hospital of a new technology.		
Technology Impact	Patient < 65 yrs (Private insurance – per diem)	Patient > 65 yrs (DRG – Medicare)
Decrease LOS	Not attractive	Very attractive
Decrease cost/day	Not attractive	Very attractive

Thus, it becomes very important to know the patient "mix" — private vs. Medicare/Medicaid — that a particular hospitals serves. For instance, a hospital with 80% Medicare/Medicaid patients will be very interested in new technology that will reduce LOS and/or cost/day. Hospitals with 80% private insurance will not necessarily see the benefit to reducing LOS and/or cost/day. All hospitals know and keep track of their mix of patients for this reason. It is very important to document the patient mix when completing a 3Q analysis and making recommendations. Note that even though a hospital with 80% private insurance may not be attracted to new technology, they may see the new technology as a marketing advantage in a competitive arena and choose to purchase the technology for that reason.

APPENDIX C

Technology Assessment PowerPoint Report Guidelines

Figure C.1:

1. Remember to whom you are giving the report. Test by pretending you are your manager and read it through his/her eyes. Is the language and detail appropriate for him/her?

2. The less information on a slide, the more visual impact the slide will have. It's fine to reveal (or "build") a concept over multiple slides. Graphics are the clearest way to present statistics.

3. Include statistics from research sources to give credibility to your conclusions and recommendations, but don't make them too complex.

4. Use slide titles to more effectively communicate the exact nature of the slide. Try this test: After you think you have the final slide report finished, put it in "slide sorter" view, big enough so you can read the slide titles. Then read only the slide titles all the way through the presentation and see if they communicate the specific flow and information of your presentation. If they don't, then reorganize the slides or modify the titles.

5. If you are using a slide title more than once, use "continued" at the top of the slide.

6. Document all references on your slides, preferably at the bottom of the slide. This includes telephone conversations or e-mails you have had with experts, vendors, etc., for example: Personal communication with Dr. Harold Smith, Cornell University Medical Center, February 1, 2009.

7. Your slide report must be able to *stand by itself.* Your manager may use it without you being present, so you have to include all critical information on your slides or in the Notes portion of the slide. You also want your manager to feel comfortable taking your presentation to the next level of the organization.

A P P E N D I X   D

Class Report Scenario Example

Report #1:   Assigned 3/25/09
Presented 4/8/09

TECHNOLOGY TO BE ASSESSED:

Clarian GlucoStabilizer^{TM} - computerized intravenous insulin infusion program to control blood glucose in the ICU.

SITUATION:

You are a Technology & Business Development Specialist in the *Solar Medical Company*. Your manager, VP of Business Development, has asked you and a team member to assess this technology using the 3Q Method for possible licensing and integration into Solar's patient monitoring product line. *Clear and actionable recommendations are required.*

Your assessment report must be completed and presented at a monthly review meeting, April 8th. Attending this meeting will be a) your immediate manager, b) the VP of Marketing, and c) the company CFO.

DIRECTIONS:

1. Use all 3 parts of the 3Q method.

2. Besides the main presentation, create one four-blocker summarizing your findings and recommendation(s).

3. Keep the presentation to 20-25 minutes.

APPENDIX E

Four-Blocker Slide Templates for 3Q Reports

3 Question Assessment –
4 Blockers
(Company Version)

P. Weinfurt

HCTM 210 Case Analysis **[Team Names]**	IS IT REAL?	[Company] [Date] [Technology]

Solve a Real Clinical Problem?

1. What clinical problem does it solve?
2. Projected clinical impact?
3. Create any new problems?

TAKEAWAY

Clinically Tested?

1. Valid study? (MAARIE or Gold standard comparison studies)
2. Clinical limits tested?
3. Is it clinically usable?
4. How complex is training?

TAKEAWAY

Based Upon Real Science?

1. Principles involved?
2. Research/clinical literature?

TAKEAWAY

Summary

TAKEAWAY

HCTM 210 Case Analysis [Team Names]	CAN WE WIN?	[Company] [Date] [Technology]

Strategic fit?

1. Market growth > Z%?
2. Will it improve our competitiveness?
3. Give us leadership image?
4. Will it leverage sales of existing products?
5. Will it cannibalize existing products?

Takeaway

Have WW Distribution Channels?

1. Do we presently sell to target customer?
2. Product include disposables?
3. Can we service it?
4. Use present training staff?

Takeaway

Can We Be Competitive?

1. Who are competitors?
2. What are competitors market share?
3. Can we obtain X% market share within Y years?

Takeaway

Customer Acceptance Curve?

1. Does it replace an existing product or technology?
2. How large a marketing & education effort to move customer to this technology?

Takeaway

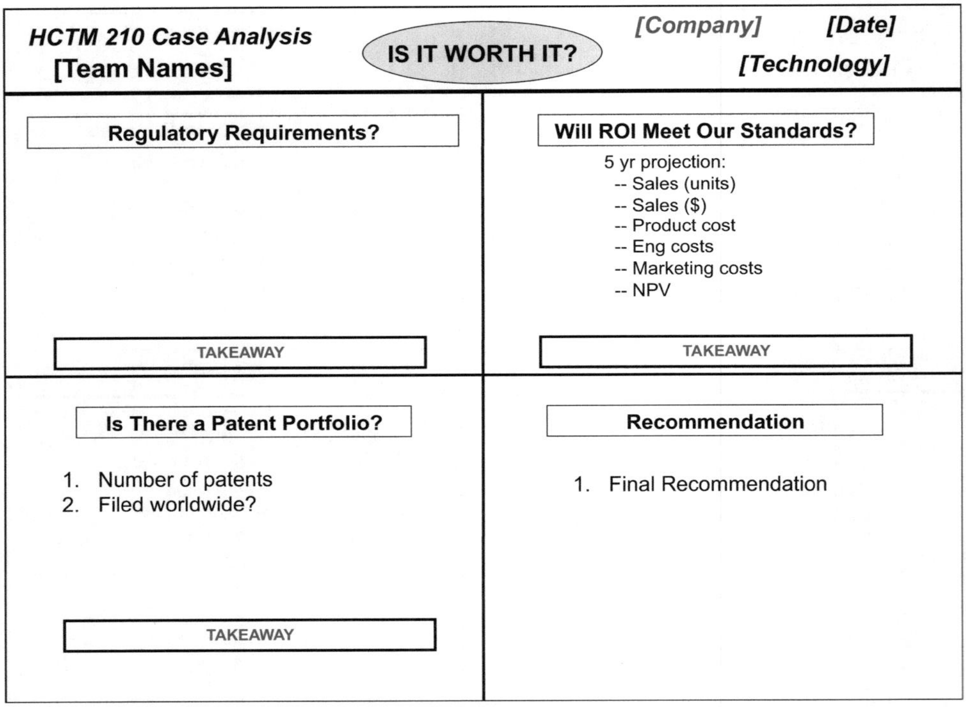
HCTM 210 Case Analysis
[Team Names]
IS IT WORTH IT?
[Company]
[Date]
[Technology]
Regulatory Requirements?
TAKEAWAY
Will ROI Meet Our Standards?
5 yr projection:
-- Sales (units)
-- Sales ($)
-- Product cost
-- Eng costs
-- Marketing costs
-- NPV
TAKEAWAY
Is There a Patent Portfolio?
1. Number of patents
2. Filed worldwide?
TAKEAWAY
Recommendation
1. Final Recommendation

3 Question Assessment – 4 Blockers
(Hospital Version)

P. Weinfurt

HCTM 210 Case Analysis **[Team Names]**	IS IT REAL?	*[Hospital]* *[Date]* *[Technology]*

Solve a Real Clinical Problem?

1. What clinical problem does it solve?
2. Projected clinical impact?
3. Create any new problems?

TAKEAWAY

Clinically Tested?

1. Valid study? (MAARIE or Gold standard comparison studies)
2. Clinical limits tested?
3. Is it clinically usable?
4. How complex is training?

TAKEAWAY

Based Upon Real Science?

1. Principles involved?
2. Research/clinical literature?

TAKEAWAY

Summary

TAKEAWAY

HCTM 210 Case Analysis **[Team Names]**	CAN WE WIN?	[Hospital] [Date] [Technology]

Strategic Fit?

1. Impact on competitiveness and market share?
2. Give us leadership image?
3. Is it synergistic with our existing services & markets?
4. Safety and risk to patients and health care workers?
5. Potential for standardization?
6. Will it cannibalize existing service?
7. Requirements for facility modification & work flow

Takeaway

Vendor Capability?

1. Vendor's reputation?
2. Vendor capacity to provide sufficient & reliable service & supplies?

Takeaway

Can We Be Competitive?

1. Who are the competitors?
2. What are competitors market shares?
3. Can we increase our market share?

Takeaway

Summary

Takeaway

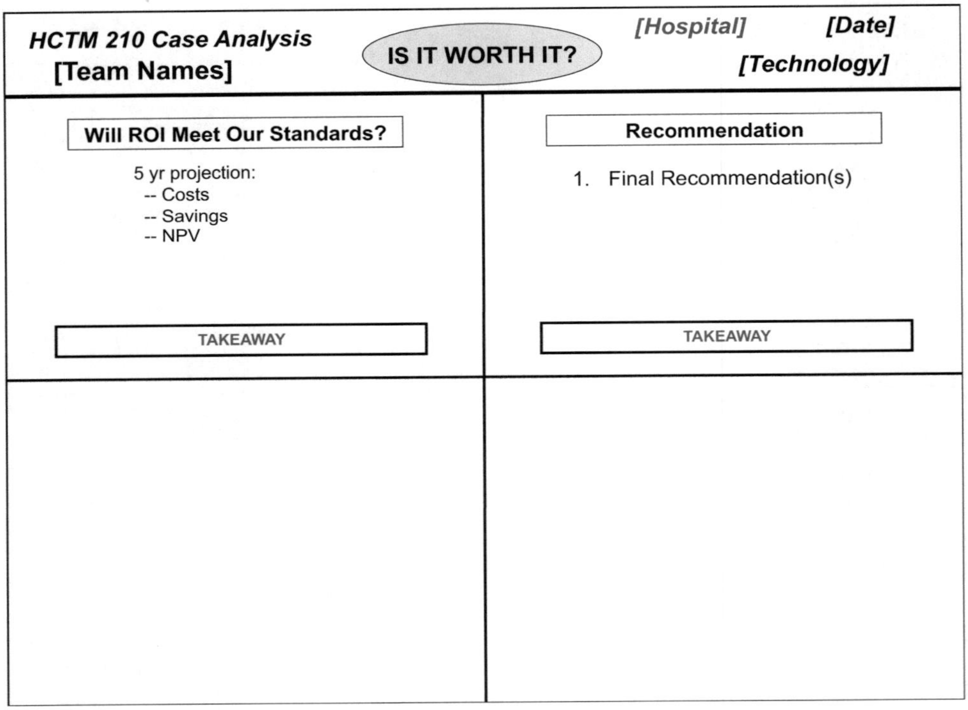
HCTM 210 Case Analysis
[Team Names]
IS IT WORTH IT?
[Hospital]
[Date]
[Technology]
Will ROI Meet Our Standards?
5 yr projection:
-- Costs
-- Savings
-- NPV
TAKEAWAY
Recommendation
1. Final Recommendation(s)
TAKEAWAY

Author's Biography

PHILLIP WEINFURT

Dr. Weinfurt is a faculty member in the Healthcare Technology Management program jointly given by Marquette University and the Medical College of Wisconsin, Milwaukee, Wisconsin. He received his bachelor's and master's degrees in electrical engineering and Ph.D. in Biomedical Engineering. Prior to becoming a faculty member at Marquette/Medical College of Wisconsin, Dr. Weinfurt spent 31 years in the healthcare technology assessment, business development, medical device and medical information systems development at GE Healthcare and Marquette Medical Systems. For 13 years while working for industry, he also conducted strategic planning regarding issues of technology and quality for the Society of Critical Care Medicine. Dr. Weinfurt has also served as a private consultant for industry for the past 8 years.

Printed in the United States
by Baker & Taylor Publisher Services